a LANGE medical book

Cell Physiology

David Landowne, PhD

*Professor
Department of Physiology and Biophysics
University of Miami
Leonard M. Miller School of Medicine
Miami, Florida*

Lange Medical Books/McGraw-Hill
Medical Publishing Division

New York Chicago San Francisco Lisbon London Madrid Mexico City
Milan New Delhi San Juan Seoul Singapore Sydney Toronto

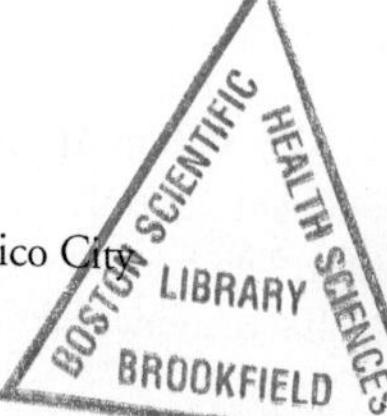

The McGraw-Hill Companies

Cell Physiology

1 2 3 4 5 6 7 8 9 0 DOC/DOC 0 9 8 7 6

ISBN: 0-07-146474-3
ISSN: 1558-9277

Notice

Medicine is an ever-changing science. As new research and clinical experience broaden our knowledge, changes in treatment and drug therapy are required. The author and the publisher of this work have checked with sources believed to be reliable in their efforts to provide information that is complete and generally in accord with the standards accepted at the time of publication. However, in view of the possibility of human error or changes in medical sciences, neither the author nor the publisher nor any other party who has been involved in the preparation or publication of this work warrants that the information contained herein is in every respect accurate or complete, and they disclaim all responsibility for any errors or omissions or for the results obtained from use of the information contained in this work. Readers are encouraged to confirm the information contained herein with other sources. For example and in particular, readers are advised to check the product information sheet included in the package of each drug they plan to administer to be certain that the information contained in this work is accurate and that changes have not been made in the recommended dose or in the contraindications for administration. This recommendation is of particular importance in connection with new or infrequently used drugs.

This book was set in Adobe Garamond by International Typesetting and Composition.
The editors were Jason Malley, Robert Pancotti, and Lester A. Sheinis.
The production supervisor was Sherri Souffrance.
The text designer was Eve Siegel.
The cover designer was Mary McKeon.
The illustration coordinator was Maria T. Magtoto.
The indexer was Alexandra Nickerson.
RR Donnelley was printer and binder.

This book is printed on acid-free paper.

INTERNATIONAL EDITION ISBN: 0071104860

Contents

Preface

The purpose of this book is to introduce readers to cell physiology in a practical manner. The book was written with three purposes in mind: as an introductory text for medical students, as an underlying basis for the study of physiology of the various organ systems of the human body, and as a review for examinations for medical licensure. In addition, the book may be useful as a survey of cell physiology and membrane biophysics for beginning graduate students in physiology, who then may pursue a chosen area of research in greater depth. I hope that the book is also helpful for students of allied health and nursing and for researchers in health-related fields who seek an introduction to cell physiology.

I have attempted to write this book in a way that is comprehensible and that facilitates a working knowledge of the material. With appropriate support, first-year medical students can learn this material in a two-week intensive course. Each chapter includes a set of study questions and a list of suggestions for further reading. An NBME-style practice examination is provided as an appendix.

As described in the first chapter, feedback is an important component of any organized activity. I welcome any questions, suggestions, and corrections that will make this book more useful. Please write to me directly at *dl@miami.edu.*

I want to thank my colleagues and students who have shown me the aspects of cell physiology that are important from a medical perspective. I want to acknowledge the editors at McGraw-Hill for their assistance and the director and staff of the Marine Biological Laboratory in Woods Hole, Massachusetts, for their excellent library. Special thanks to my father, Milton, my wife, Edith, and my children, Mahayana and Youme.

To J. F. Danielli and A. C. Giese.

Cellular Processes 1

OBJECTIVES

- *Recognize and describe the types of electrophysiological events.*
- *Describe the types of membrane channels and their roles.*
- *Describe physiological control systems.*

OVERVIEW

Physiology is the study of functions or processes. Diogenes Laertius in his *Lives of Eminent Philosophers* declared that there are three divisions of philosophy: natural, ethical, and dialectic. The ancient Greek word for natural philosophy was φυσισ, which is the root for the English words *physics*, *physiology*, and *physician*. Physics and physiology are about how things work. The practice of medicine is the physician's job; physiology is the scientific basis for that practice.

Life is cellular, and cells are the fundamental units of life. Without cells there would be no living beings. All the cells of a given individual are ultimately derived from a single fertilized ovum. Most of the cells of multicellular organisms reside within their tissues and organs. This book concentrates on the cellular processes and leaves the discussion of their higher organization to works on the physiology of the various organ systems. Drugs, toxins, and diseases are introduced to illustrate the cellular processes. Other books will be needed to understand these in the context of medicine. A physician's patients are more than their cellular physiology, but the quality of their lives depends on their cellular functionality.

COMMUNICATION

This book is about the dynamic cell processes that support sensory perception of the environment, communication, and the integration of information within and between cells as well as their expression, or actions on the environment. These are the processes that enable the cell to contribute to the functioning of tissues, organs, and individuals. These processes make up Norbert Wiener's third fundamental phenomenon of life, which he referred to as *irritability* and is now usually called *excitability*. The other two phenomena of life, reproduction and metabolism, also

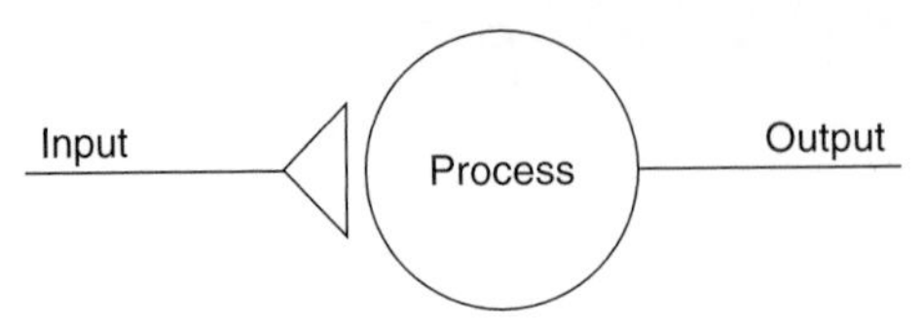

Figure 1-1. The input-process-output structural framework is a specification of causal relationships in a system.

occur in all cells but are not covered in depth here. The consideration of perception, integration, and expression can be generalized to the consideration of physiological events in terms of inputs, processes, and outputs (Fig. 1-1). Complex processes can be broken down into simpler ones, with the outputs of one or more processes becoming the inputs to the next one.

In order to survey the processes discussed here, it is useful to consider a three-cell model of the body. Figure 1-2 shows a sensory **neuron** or nerve cell, a motor neuron, and a skeletal muscle cell. These cells represent the hardware the body uses to carry out these functions. The cells have specialized portions for the different processes. Starting from the left, the sensory cell has one end specialized for the transduction of a stimulus into a cellular signal. The various senses have different specializations here to accomplish this transduction. Besides the classic five senses (touch, hearing, vision, taste, and smell), there are sensors or proprioceptors inside the body that sense internal parameters—e.g., body temperature, blood pressure, blood oxygen levels, or the lengths of various muscles.

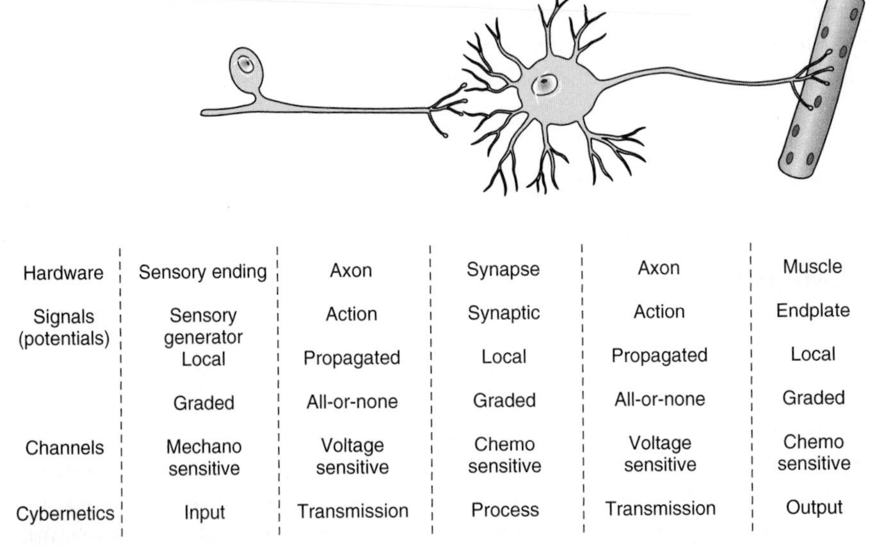

Hardware	Sensory ending	Axon	Synapse	Axon	Muscle
Signals (potentials)	Sensory generator Local	Action Propagated	Synaptic Local	Action Propagated	Endplate Local
	Graded	All-or-none	Graded	All-or-none	Graded
Channels	Mechano sensitive	Voltage sensitive	Chemo sensitive	Voltage sensitive	Chemo sensitive
Cybernetics	Input	Transmission	Process	Transmission	Output

Figure 1-2. The cellular processes of a hypothetical three-celled organism.

1 If it is sufficiently large, the initial signal causes another signal to propagate over the **axon** (the long cylindrical portion of the nerve cell) until it reaches the other end, where the sensory neuron makes a synaptic connection with dendrites of the motor neuron, located in the central nervous system (CNS). The message is transmitted from the **presynaptic** cell to the **postsynaptic** cell, where it is integrated or evaluated along with messages from other neurons that synapse on the same motor neuron. In the complete organism, this integration and comparison occurs in many cells and at different levels within the CNS, so the decision to move or not move can be made considering more than one input and also anything the organism has learned from the past.

If the motor neuron is sufficiently excited, it will send another message along the axon that leads to a synapse on a muscle cell. In healthy people, this neuromuscular synapse always leads to a signal that propagates over the length of the muscle cell and activates contraction, which can act on the environment. Other actions on the environment are effected by the secretions of various glands; these too may be controlled by synaptic connections. These muscles and glands may act internally (e.g., to control heart rate or blood pressure) or externally (for locomotion or communication with other people).

These signals are all electrical; they all represent changes in the electrical potential difference across the various cell membranes. Every living cell has a surface membrane that separates its intracellular and extracellular spaces. All cells, not just those of nerve and muscle, are electrically negative inside the cell with respect to outside. This is called the **membrane potential**. When the cells are "resting"—that is, not signaling—their membrane potential is called the **resting potential**. Chapter 3 is about the origins of the resting potential.

2 Even though the signals described above are changes in potential, they are generally referred to as *named potentials*. On the left, there is the **sensory generator potential**, which has two properties to distinguish it from the next signal, the **action potential**. The sensory generator potential is **local**; it is seen only within a few millimeters of the sensory ending. The action potential is **propagated**; it travels from the sensory ending to the presynaptic terminal, perhaps more than a meter away. The sensory generator potential is also **graded**; a larger-amplitude stimulus produces a larger-amplitude sensory generator potential. In contrast, the action potential has a stereotyped amplitude and duration; it is **all-or-none**. The information about the stimulus is encoded in the number of action potentials, or the number per second. A larger-amplitude stimulus will result in a higher frequency of action potentials, each with the same stereotyped amplitude. Because the all-or-none character of neurons is similar to the true-or-false character of logical propositions, cyberneticists have considered that neural events and the relations among them can be treated by means of propositional logic. Chapters 4 and 5 are about sensory generator potentials and action potentials, respectively.

3 The presynaptic terminals contain a mechanism to release the contents of **vesicles** containing chemical **transmitters** that diffuse across the narrow **synaptic cleft** and react with the postsynaptic cell to produce a **postsynaptic potential**. The postsynaptic potential is also local and graded. It is only seen within

a few millimeters of the site of the presynaptic ending and its amplitude depends on how much transmitter is released. There are **excitatory postsynaptic potentials (EPSPs)** and **inhibitory postsynaptic potentials (IPSPs)**, depending on whether the postsynaptic potential makes the cell more or less likely to initiate an action potential. If there is sufficient excitation to overwhelm any inhibition that may be occurring, an action potential will be initiated in the postsynaptic cell. There are many presynaptic cells ending on each postsynaptic neuron as well as various different transmitters in different synapses. These transmitters, the release mechanism, and the resulting postsynaptic potentials are discussed in Chap. 6.

The action potential in the motor neuron and the synapse with the muscle cell are very similar to the previous cases. In the light microscope, the neuromuscular junction looks like a small plate; hence the junction is often called an endplate and the postsynaptic potential an **endplate potential**. The neuromuscular junction differs from most other synapses because there is only one presynaptic cell, its effect is always excitatory, and—in healthy people—is always large enough to initiate an action potential in the muscle cell.

The muscle action potential propagates along the length of the cell and into the interior by small **transverse tubules**, whose membranes are continuous with the surface membrane. The action potential excitation is coupled to the muscular contraction by processes described in Chap. 7. That chapter also discusses the control of cardiac and smooth muscle cells.

The resting potential, the sensory generator potentials, the action potentials, and the synaptic potentials all occur by the opening and closing of **channels** in the cell membranes. These channels are made of proteins that are embedded in and span the membrane connecting the intracellular and extracellular spaces. Each has a small pore through the middle, which may be opened or closed and is large enough to allow specific ions to flow through and small enough to keep metabolites and proteins from flowing out of the cell. There are many channels, and a good part of Chap. 2 is devoted to their description. They are generally named either for the ion that passes through them or for the agent that causes them to open.

(4) There are three classes of channels that act to produce the changes in potential shown in Fig. 1-2. All these channels will be discussed individually in Chap. 2 and then again in the context of the various potentials in the rest of the book.

Mechanosensitive channels subserve the sensations of touch and hearing and the many proprioceptors that provide information on muscle length, muscle tension, joint position, the orientation and angular acceleration of the head, and blood pressure. These channels open when the membrane of the sensory ending is stretched, sodium ions flow through the channels, and the membrane potential changes.

Voltage-sensitive channels underlie action potentials. They open in response to a change in membrane potential. When they are open, ions flow through them, and this changes the membrane potential as well. The generator potential or the synaptic potentials start these channels, and then they open the remaining adjacent voltage-sensitive channels. This accounts for the propagation and all-or-none, stereotyped quality of the action potentials. Nerve and skeletal muscle action potentials are produced by the successive activation of voltage-sensitive sodium channels, followed by voltage-sensitive potassium channels. There are also voltage-sensitive calcium channels in the presynaptic nerve endings. When the action potential reaches the presynaptic terminal, these calcium channels open and permit calcium to enter the cell. The calcium binds to intracellular components and initiates the release of synaptic transmitters.

Chemosensitive channels are responsible for the synaptic potentials. The transmitters bind to these channels, causing them to open. There are different channels for different transmitters and also different channels for EPSPs and IPSPs. Chemosensitive channels also subserve the chemical senses of smell and taste. There are also channels that open or close in response to intracellular chemicals such as adenosine triphosphate (ATP) or the cyclic nucleotides, cyclic adenosine monophosphate (cAMP) or cyclic guanosine monophosphate (cGMP). Vision is supported by a reaction series whereby light absorption leads to a decrease in cGMP, which produces a closure of cyclic nucleotide–gated (chemosensitive) channels. When sodium ions stop flowing through these channels, the membrane potential changes.

From a cybernetic viewpoint, Fig. 1-2 indicates that the body has mechanisms to input information, to transmit it within the body, to process the information, and to provide output. This type of analysis appears frequently in physiology. Much of what you will learn here can be broken into various steps where the output of one process becomes the input for the next. For example, the sensory generator potentials are an input to the action potential–generation process and the action potential is the input to the voltage-sensitive calcium channel, which permits calcium to enter the presynaptic terminal. This calcium is the input for the transmitter release process, and so on.

CONTROL

Although most of this book focuses on isolating the different processes so as to analyze them more easily, an understanding of the value and true significance of each physiological quality must refer to the whole organism. A recurring theme throughout all of physiology is the maintenance of a stable internal environment through **homeostasis**. Many internal properties (e.g., body temperature or blood glucose levels) are homeostatically controlled within narrow limits by feedback control systems.

5 Homeostasis is a property of many complex open systems. Feedback control is the central feature of organized activity. A homeostatic system (e.g., a cell, the body, an ecosystem) is an open system that maintains itself by controlling many dynamic equilibria. The system maintains its internal balance by reacting to changes in the environment with responses of opposite direction to those that created the disturbance. The balance is maintained by **negative feedback**.

Perhaps the most familiar negative feedback control system is the thermostat that controls the temperature of a room or house. This device measures the temperature and compares it to a set point, the temperature that is desired. If the actual temperature is colder than desired, a signal is sent to send in some heat, perhaps by turning on a heater. If the actual temperature is too hot, the heater will be turned off and an air conditioner will be turned on. The control of body temperature uses muscle contraction, or shivering, to raise the temperature and perspiration and its evaporation to lower the temperature.

The basic steps (Fig. 1-3*A*) in negative feedback control of any measurable parameter are the measurement by a **sensor, communication** of that measurement to a **comparator**, making the comparison, and communicating the comparison to an **effector** that changes the parameter of interest. The feedback is called negative because the signal to the effector reduces the difference between the measured value and the desired value. In the case of the thermostat, this can be done either by supplying heat or by removing it, depending on the need at the time.

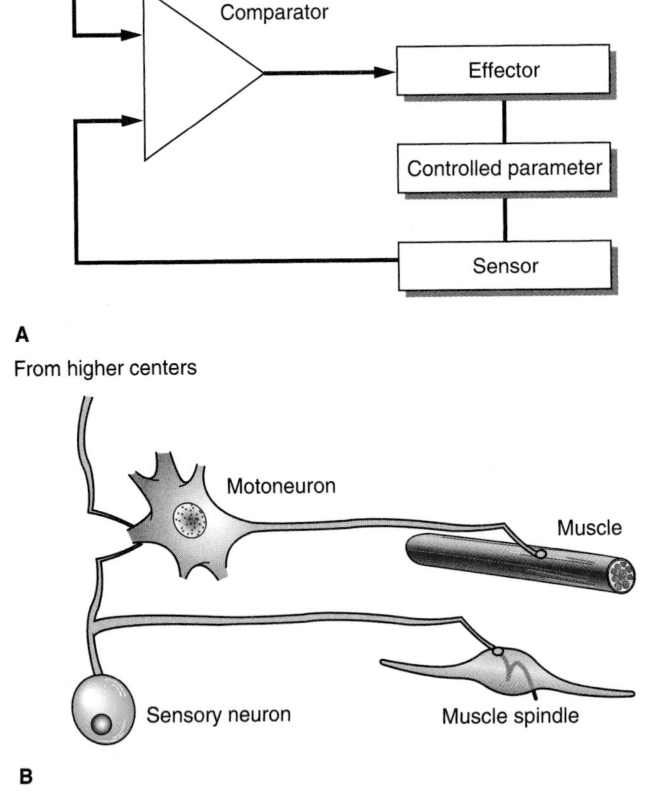

Figure 1-3. Homeostasis and feedback control.

The three cells in Fig. 1-2, arranged as a negative feedback loop (Fig. 1-3*B*), represent the process used to control the length of muscles both to maintain posture and to achieve movement in response to signals from the brain. This feedback loop can be easily demonstrated by the **stretch reflex**—i.e., the knee-jerk reflex. If the patellar tendon is tapped, the quadriceps muscles will be stretched, and this will be detected by stretch receptors embedded in the muscle. Mechanosensitive channels will open changing membrane potentials in those sensory endings that will induce action potentials to propagate through the dorsal roots and into the nerve terminals in the ventral horn of the spinal cord. Transmitter will be released, which will excite the nerve leading from the spinal cord out the ventral roots and back to the quadriceps muscle cells, where the synaptic process will be repeated and the muscle will shorten to compensate for the initial stretch. When the organism wants to change the length of the quadriceps, the signal from the brain may be sent through a cell in the spinal cord near the motor neuron that changes the desired length for that particular muscle. There are also higher-order feedback systems that control organized contraction of many muscles to achieve complex behaviors such as walking.

You will meet many negative feedback control systems as you study physiology. There are also a few **positive feedback** systems of which it is well to be aware. A positive feedback system is unstable; the signal from the sensor increases the effect, which increases the signal from the sensor in a "vicious cycle," which is limited only by the availability of resources. Take, for example, an explosion, where heat ignites a chemical, which produces heat, which then ignites more chemical until all the chemical is consumed. The upstroke of the action potential is governed by a positive feedback loop; this accounts for the all-or-none property of action potentials.

Your study of physiology will be easier if you recognize the many examples of negative feedback loops and identify the sensor, the comparator, the effector, and the communication pathways, which may be neuronal, hormonal, or cellular. The physician is aided by an understanding of these homeostatic mechanisms by considering impairments in feedback control. Appropriate treatment for the loss of control will depend on which part of the feedback loop has been compromised.

KEY CONCEPTS

Communication in excitable cells occurs via electrical signals within the cells and via chemical signals at synapses between the cells.

There are two classes of electrical signals: those that are local and graded and those that are propagated and stereotyped, or all-or-none.

The chemical transmitters are released presynaptically and produce an electrical signal in the postsynaptic cell.

Three classes of ion channels produce the electrical signals: mechanosensitive, chemosensitive, and voltage-sensitive channels.

Homeostasis by negative feedback control is an important feature of living systems.

There are three basic elements of a negative feedback loop: a sensor, a comparator, an effector, and two communication links connecting them.

STUDY QUESTIONS

1–1. Name three different electrical signals in body cells. For each signal, describe two distinguishing qualities and the class of membrane channels that produce the signal.

1–2. Draw a negative feedback loop and label the components.

SUGGESTED READINGS

Diogenes Laertius. *Lives of Eminent Philosophers.* Trans. R.D. Hicks. Cambridge, MA: Harvard University Press, 1990. http://classicpersuasion.org/pw/diogenes/

Wiener, Norbert. *Cybernetics, or Control and Communication in the Animal and Machine,* 2d ed. Cambridge, MA: MIT Press, 1965.

Cell Membranes 2

OBJECTIVES

- *Describe the molecular composition of biological membranes.*
- *Describe the functional biophysical properties of biological membranes.*
- *Describe classes of ion channels, their molecular structure, and their biophysical properties.*
- *Describe the molecular organization, properties, control, and functional roles of cell-cell channels.*
- *Describe the movement and transport of substance across biological membranes by passive processes.*
- *Describe the movement and transport of substance across biological membranes by active processes.*
- *Describe the physiological importance of two examples of active and two examples of passive transport.*
- *Define osmotic pressure.*
- *Calculate the osmolarity of simple solutions.*
- *Calculate the changes in osmolarity in body compartments caused by drinking various simple solutions.*
- *Describe physiological mechanisms to regulate osmolarity.*

Every living cell has a surface membrane that defines its limits and the connectivity of the intracellular and extracellular compartments. Cell membranes are about 10 nm thick and consist of a 3- to 4-nm-thick **lipid bilayer** with various embedded proteins that may protrude into either compartment. Membranes also delimit intracellular organelles, including the nuclear envelope, Golgi apparatus, endoplasmic reticulum, mitochondria, and various vesicles. The proteins handle the transport of specific molecules across the membranes and thus control the different solutions on either side. The proteins also support communication across the membranes and along the surface of the cell. There are also proteins that provide mechanical coupling between cells.

LIPIDS

Most of the membrane lipids are **glycerophospholipids**, which have a glycerol backbone with two of its three –OH groups esterified by fatty acids and the third esterified to a phosphate group, which is in turn esterified to a small molecule that gives its name to the whole molecule (Fig. 2-1). The most common glycerophospholipids are **phosphatidylcholine** (PC), **phosphatidylethanolamine** (PE), and **phosphatidylserine** (PS). Membranes also contain **phosphatidylinositol** (PI), which plays an important role in signaling within the cytoplasm. Notice that PS and PI have a net negative charge. Animal-cell membranes may also contain sphingolipids, including the phosphosphingolipid, **sphingomyelin**, which has two acyl chains and a phosphate-linked choline head linked to a serine backbone, and glycosphingolipids, which have sugars in the head group. Membranes also contain **cholesterol**, which has a steroid ring structure.

All of these lipids are **amphipathic** because they have **hydrophilic**, or "water-loving," head groups and **hydrophobic**, or "water-fearing," acyl tails. The –OH group of cholesterol is hydrophilic and the rest is hydrophobic. A hydrophobic effect arises from the lack of interactions of hydrocarbons with water and the strong attraction of water for itself. Thus, when placed in an aqueous environment, these lipids spontaneously assemble into closed bilayer membrane

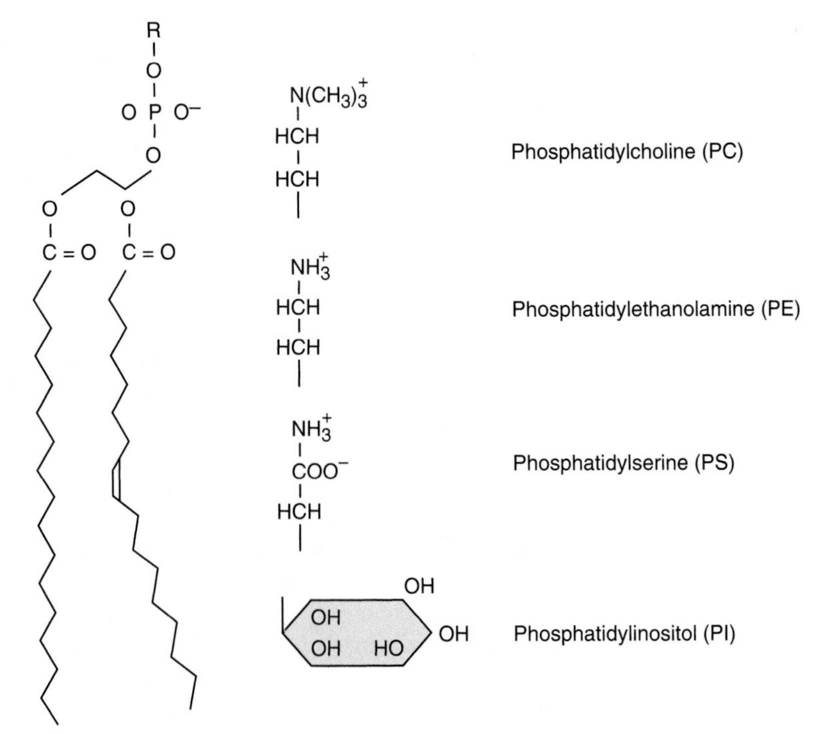

Figure 2-1. Glycerophospholipids.

vesicles. Detergents are also amphipathic molecules; however, since they have only a single acyl chain, detergents assemble as micelles or spheres with the hydrophobic tails inside. Detergents can be used to destroy lipid membranes and to extract proteins that were embedded in the lipids.

The lipids are relatively free to diffuse laterally within the plane of the membranes, but—with the exception of cholesterol—they are unlikely to **flip-flop** from one half of the bilayer to the other owing to the hydrophobicity of the head groups. The bilayer is asymmetrical, with the choline-containing phospholipids, PC and sphingomyelin, in the outer half and the amino-containing phospholipids, PE and PS, in the inner half. In addition, the glycosphingolipids are in the noncytoplasmic half and PI is facing the cytoplasm. The asymmetrical arrangement is produced as the membranes are assembled in the endoplasmic reticulum. The phospholipids are synthesized and inserted on the cytoplasmic side of the membrane; then a **phospholipid translocator** or **"flippase"** transfers PC to the noncytoplasmic side. Sphingomyelin and the glycosphingolipids are produced in the Golgi apparatus on the noncytoplasmic side.

The ease of lateral diffusion, or membrane fluidity, is increased by the presence of unsaturation or double bonds in the hydrocarbon tails. This forms a kink in the tail and therefore looser packing. At the concentrations generally found in biological membranes, cholesterol reduces the fluidity because of its rigid ring structure. Glycosphingolipid head groups tend to associate with each other and reduce fluidity. Lipid protein interactions may also reduce fluidity. There are cholesterol–sphingolipid microdomains, or **"lipid rafts,"** involved in intracellular trafficking of proteins and lipids.

PROTEINS

3 The **intrinsic proteins** of the membrane support the selective movement of ions and small molecules from one side of the membrane to the other, sense a **ligand** on one side of the membrane and transmit a signal to the other side, and provide mechanical linkage for other proteins on either side of the membrane. The proteins that move materials across the membrane can be functionally divided into **channels**, **pumps**, and **transporters**. Channels may be specific and may open and close, but, when open, they facilitate the movement of materials only down their **electrochemical gradients**. Ion channels control the flow of electrical current through the membrane. Pumps move ions up their electrochemical gradient at the expense of consuming ATP. The pumps maintain the gradients that allow the channels and transporters to do their jobs. Transporters can link the movement of two (or more) substances and can move one of them up its gradient at the expense of moving the other one down.

A protein is the product of translating a gene; it is a folded, linked sequence of alpha amino acids chosen from a palette with 20 possible different side chains. The peptide link between amino acids –CO–NH– has a planar transconformation; the folding occurs according to the torsion angles between the amino group and the alpha carbon (ϕ) and between the alpha carbon and the carboxyl group (ψ).

Hydrogen bonding between the carbonyl oxygen of a link and the fourth subsequent amino hydrogen favors a right-handed alpha-helical secondary structure with 3.6 residues per turn, a backbone diameter of about 0.6 nm, and a translation along the helix axis of 0.15 nm per residue or a pitch of 0.54 nm. When viewed with the amino terminus at the top, all the carbonyl groups point up and all the amino groups point down. The side chains point out from the helix.

A more extended secondary structure, the beta sheet, can be stabilized by hydrogen bonds between alternate carboxy and amino groups on separate strands. Each strand is a pleated sheet with a displacement of 0.35 nm per residue. The carbonyl groups point perpendicular to the strand axes, connecting the strands, and the residues point perpendicular to the sheet on alternate sides for each residue.

The conformation or tertiary structure of the entire protein is the three-dimensional relationship of all its atoms. Proteins have regions of various secondary structure connected by linkers with less easily characterized structure. Most of the proteins discussed in this book have more than one conformation. For example, a channel may be open or closed. The local secondary structures do not change very much during these conformational changes; rather, change occurs in the relationship between larger portions of the molecule.

There is also a supermolecular or quaternary level of organization. Some channels are made of a single polypeptide chain, while others are made of four to six chains. Many channels also have accessory proteins that modulate their function. In addition, the lipid matrix imposes structural restrictions on the embedded proteins.

In general proteins are amphipathic and have regions that are more hydrophobic or hydrophilic, depending on the nature of the side chains. The membrane proteins discussed here have one or more **transmembrane** (TM) alpha-helical segments with hydrophobic side chains in contact with the hydrocarbon of the lipid. If more than one helix is involved, it is possible to have hydrophobic residues facing the lipid and other groups facing each other in the more interior parts of the protein. The general pattern is for the protein to cross the membrane several times, with intracellular and extracellular loops between TM segments. There is also an N-terminal region before the first segment and a C-terminal region after the last. The N-terminal region can be on either side, but the C-terminal region is usually cytoplasmic. Either or both terminal regions can be quite large compared to the transmembrane regions.

The transmembrane folding occurs as the protein is synthesized in the endoplasmic reticulum (ER). The noncytoplasmic portions of the protein may be glycosylated in the Golgi apparatus before being inserted in the surface membrane. Subunit assembly may also occur in the ER or Golgi apparatus.

For most membrane proteins, only the primary sequence is known. Secondary structure can be predicted by sequence analysis. The presence of putative hydrophobic helices of sufficient length is taken as a suggestion of a TM segment. A topology or pattern of loops and TM segments can be predicted; such a prediction has been tested for many proteins by preparing antibodies for the putative extracellular portions. Sequence analysis of entire genomes suggests that about 20% of the proteins contain one or more TM segments and are thus membrane proteins.

Only a few membrane proteins have been crystallized and subjected to x-ray diffraction analysis. These crystals must include lipid or detergent molecules to satisfy the hydrophobic needs of the TM segments. Most of the solved structures are of bacterial proteins that have been genetically modified to enhance crystallization. A strong sequence homology between the crystallized molecule and part of the human protein is taken to indicate that both have a similar structure.

Channels, pumps, transporters, receptors, and cell adhesion molecules come in many varieties to serve many roles. The following five sections will describe a taxonomy and the anatomy of examples of each functional class. It may be useful to return to this section while reading the later part of this chapter and those parts of the rest of the book that describe the role of these molecules in physiological processes.

CHANNELS

In the previous chapter channels were distinguished by the method of opening. These were mechanosensitive channels involved in sensory processes, voltage-sensitive channels involved in action potential propagation, and chemosensitive channels involved in synaptic transmission. There are also channels that are usually open, such as channels that maintain resting potential, water channels, and specialized cell-cell channels that connect the cytoplasm of one cell with the cytoplasm of another. This section describes some channels that support various cell processes discussed later in the book. It is not exhaustive; many channels and many classes of channels are not mentioned. This is a "golden age" for ion channels. Electrophysiology and molecular and structural biology are revealing some amazing membrane proteins.

5 Many ion channels are selective and are named according to the ion that passes through them. The first channel to be crystallized is the resting potential potassium channel, also known as the **inward rectifier** or $\mathbf{K_{ir}}$. The reason for this name is discussed in the next chapter, along with its function. K_{ir} is a tetramer with four identical subunits arranged with radial symmetry and a pore that permits ion flow at the axis (Fig. 2-2*A*). Each monomer has two TM segments with an extracellular **P loop** in between (Fig 2-2*B*; see also Fig. 2-3, segments 5 and 6). The four P loops dip back into the membrane and together form the lining of a pore that goes about one-third of the way through the membrane. This pore empties into a larger intramembranous cavity that communicates with the cytoplasmic space. The eight helices form a wall for the cavity and also surround the inserted P loops. The TM helices form a conical structure with the point toward the cytoplasm.

The selectivity of the pore for potassium ions depends on the specific amino acids forming the lining. VGYGD is the K-channel signature sequence (Fig. 2-2*C*); it has been found in K channels from more than 200 organisms. This portion of the molecule is the selectivity filter because it accepts K^+ ions and excludes other ions. The pore is lined with the carbonyl oxygen groups; these are in the same relation to each other as the oxygen of water molecules that coordinate around K^+ ions

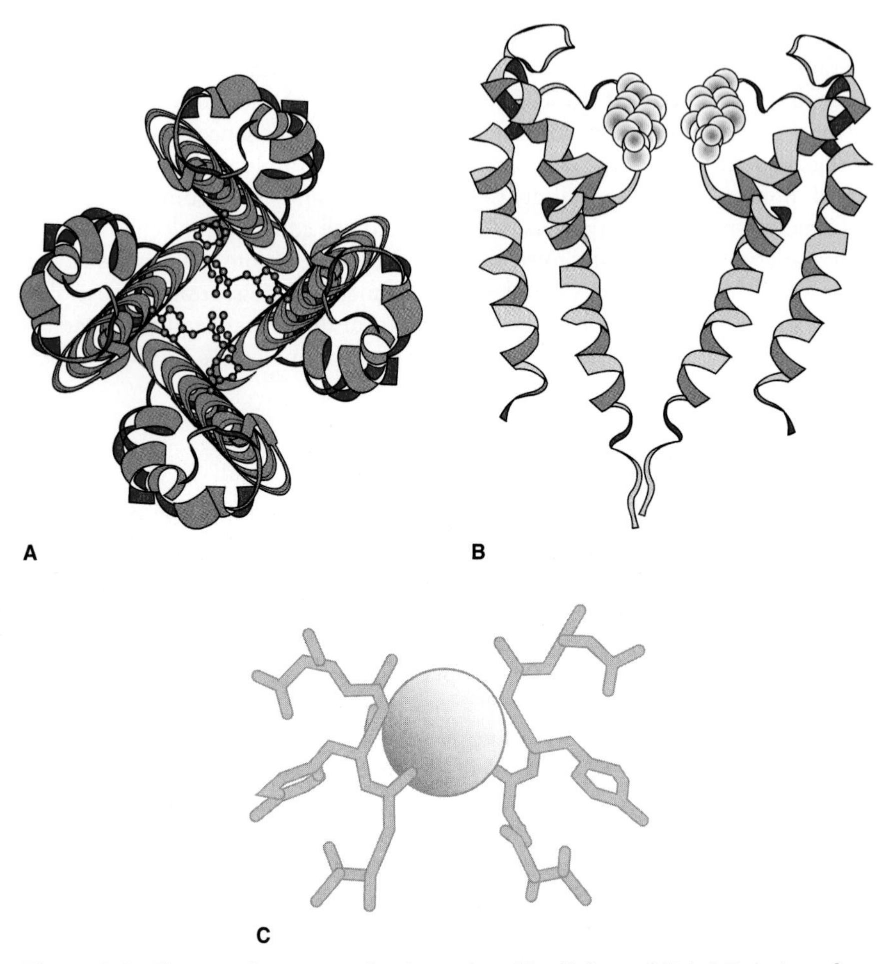

Figure 2-2. The crystal structure of an inward rectifier K channel (K_{ir}). *A.* Top view of a ribbon-structure representation with stick-and-ball for the GYG sequences (1bl8). *B.* Side view with two monomers removed; the GYG sequence is a space-filling representation (1jvm). *C.* Closeup view of two VGYGD sequences and an ion (1jvm). (Symbols in parentheses indicate Protein Data Bank identification.)

in solution because of its positive charge and the oxygen's electronegativity. Two of the coordinating oxygens from glycines just below the tyrosines can be seen in Fig. 2-2*C*. Ions with different charge or radii will coordinate water differently and thus will be less likely than K^+ ions to leave the water and enter the K channel.

It is thought that Fig. 2-2 represents a closed K_{ir} channel. The structure of another prokaryotic 2-TM channel has been solved; its inner helices are bent and splayed open, creating a wide entryway. This second K_{ir} channel responds to Ca^{2+} ions on its intracellular side by increasing its open probability. The Ca^{2+} binds to the regulator of K conductance (RCK) domain in the C-terminal part of the protein,

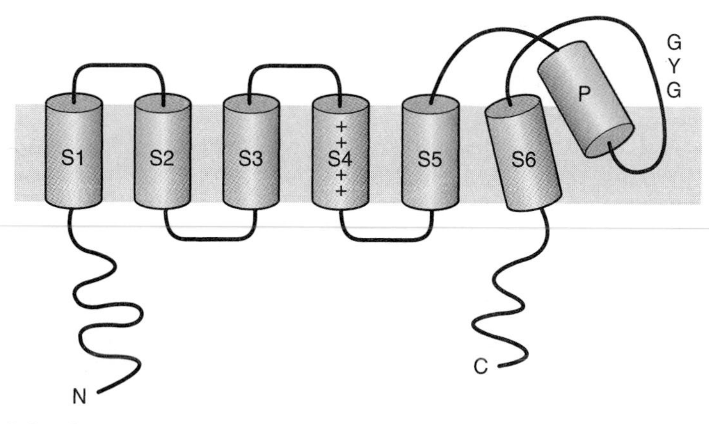

Figure 2-3. The topology of one monomer of voltage-dependent K channels (K_V).

not shown in Fig. 2-2, inducing a conformational change that splays the internal helices. Ca^{2+} and cyclic nucleotides increase the open probability of some other 2-TM and 6-TM channels by a similar mechanism.

There are eight subfamilies of 2-TM K_{ir} channels in the human genome. Several are important in cardiac electrophysiology. $K_{ir}2$ (or I_{K1}) is the original inward rectifier discovered in cardiac muscle; it is responsible for maintaining the resting potential. $K_{ir}3$ channels open via G-protein–coupled receptors; in the heart, they are referred to as K_{ACh}. $K_{ir}6$ channels open when the ADP/ATP ratio rises. In the heart, they are referred to as K_{ATP}.

Mechanosensitive Channels

Mechanosensitive channels are a diverse class of structurally unrelated channels that subserve many different functions in different cells. Mechanosensation is important for touch and hearing and also for proprioception, providing information about position, orientation, velocity, and acceleration of the body and its parts. The channels are associated with accessory molecules and cellular structures that enhance their particular functions. Somatic nonsensory cells also respond to mechanical stress without informing the nervous system—for instance, to compensate for osmotic swelling or modulate secretion or contraction. Swimming single-cell organisms can respond to hitting a barrier by reversing their direction. Bacteria have mechanosensitive channels that can respond to sudden changes in the osmotic environment and act as a safety valve.

Many mechanosensitive channels are relatively nonselective cation channels. Some are very large and permit electrolytes and small metabolites, but not proteins, to cross the membrane. The two structures that have been solved are bacterial. One is a homopentamer, with each subunit containing two TM helices. The other is a heptamer, with each subunit containing three TM helices. These are beautiful structures, but they do not shed much light on the many other forms of mechanosensitive channels.

Voltage-Sensitive Channels

Voltage-sensitive K channels (K_V) are responsible for the return to the resting state, which ends an action potential. K_V has a core structure similar to that of K_{ir} and an additional four TM helices on each subunit (see Fig. 2-3). The fourth TM segment (S4) is distinguished because it has between four and eight positively charged side chains (Arg or Lys). S4 is a signature feature of voltage-sensitive channels. It is thought to be the voltage sensor that moves toward the extracellular surface when the membrane potential changes and causes the conformational changes that lead to channel opening. There are nine subfamilies of K_V channels and several more 6-TM-channel subfamilies, including the Ca-activated K channels, the hyperpolarization-activated channels important for pacemaker activity in the heart, and cyclic nucleotide–gated channels. The last two families are nonselective cation channels.

The structure for one prokaryote K_V channel has been solved. However, there is controversy in the field over the meaning of this structure, as it seems to be at odds with many mutational electrophysiological experiments. The description in the previous paragraph is general enough to cover all sides of this controversy.

Voltage-sensitive Na channels (Na_V) are responsible for the upstroke of the action potential and support its propagation. Voltage-sensitive Ca channels (Ca_V) couple membrane potential changes with an increase in intracellular Ca concentration, which acts as a second messenger to control a variety of intracellular processes. Na_V and Ca_V channels have a structure similar to the K_V channels except they are single larger molecules incorporating four domains, each with slightly different 6-TM segments (Fig. 2-4). The selectivity filters have four different walls. The Ca_V channel has four characteristic glutamates (EEEE) in its pore lining, one on each domain. The Na_V channel has a DEKA pattern on the four walls of its pore. These side chains must be exposed to the lumen of the pore. The charges they expose to the lumen and the size of the pore determine the selectivity of the channel.

Chemosensitive Channels

There are many different chemosensitive or ligand-gated channels. These control the flow of ions and thus generate electrical signals in response to specific chemicals, such as acetylcholine (ACh), glutamate, or ATP. They can be grouped into three different superfamilies according to the stoichiometry and membrane topology of their subunits. Many of these were first discovered pharmacologically by noticing that certain compounds, called **agonists**, produced membrane currents or altered the electrical activity of cells and other compounds; **antagonists** could block these effects. For some agonist-induced currents, the ligand binds the same molecule that contains the pore. These are the ligand-gated channels, which are sometimes called **ionotropic** ligand receptors to distinguish them from **metabotropic** ligand receptors, where the ligand binds a **G protein–coupled receptor** (GPCR) and triggers a biochemical cascade that may include opening of other channels, e.g. K_{ACh}, described above.

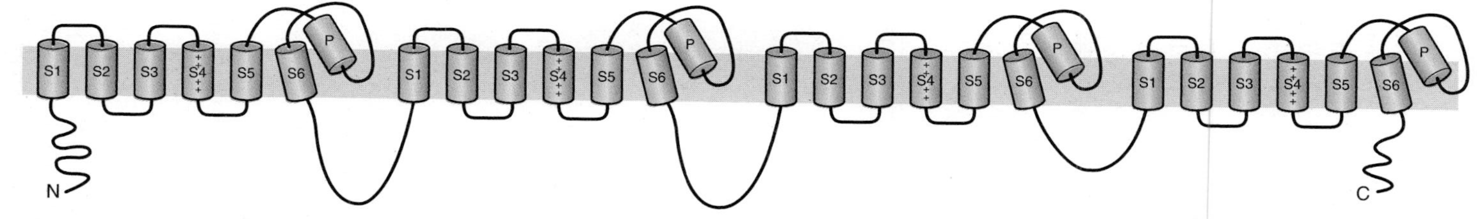

Figure 2-4. The topology of voltage-dependent Na channels (Na_V).

The ACh receptor channels (AChRs) are referred to as **nicotinic** AChRs, or nAChRs. The term *nicotinic* indicates these receptor bind nicotine, which also opens the channels. nAChRs are distinguished from **muscarinic** AChRs, which are not channels but rather GPCRs. nAChRs are found on the postsynaptic membranes at skeletal neuromuscular junctions and in the autonomic and central nervous systems.

The best-studied nAChRs are heteromeric pentamers (Fig. 2-5). The monomers have four TM segments each and a large extracellular N-terminal region. At the neuromuscular junction, the nAChR has two alpha subunits, with ACh binding sites at the interface between subunits and far from the lipid membrane. ACh binding induces a conformational change that opens the pore formed at the level of the lipid membrane and lined by the second TM segment of each of the monomers five subunits. The open channels are highly permeable to both Na and K, slightly permeable to Ca, and not permeable to anions. They are not as selective as the K_{ir} or

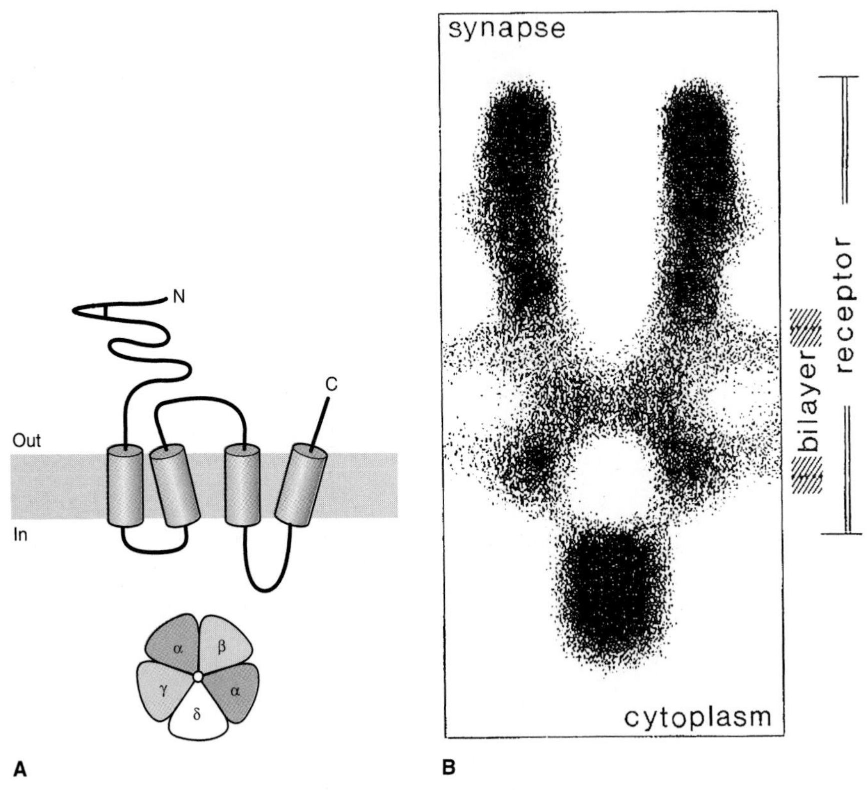

Figure 2-5. *A.* The topology of one monomer of nicotinic acetylcholine receptor channels (nAChR), with a top view showing the arrangement of the five monomers. *B.* Side view of the channel. (*B* is from Toyoshima C, Unwin N. Ion channel of acetylcholine receptor reconstructed from images of postsynaptic membranes. *Nature* 1988;336:247–250, with permission.)

voltage-sensitive channels. Functionally, the Na permeability is most important, as discussed in Chap. 6.

The CNS postsynaptic receptors for glycine (glyR), gamma-aminobutyric acid ($GABA_AR$), and serotonin ($5HT_3R$) have similar pentameric architecture, although some are homomeric, as are some nAChRs. glyRs and $GABA_ARs$ are selectively permeable to anions and produce inhibitory postsynaptic potentials (IPSPs.) $5HT_3Rs$ are cation-selective, similar to nAChRs, and produce excitatory postsynaptic potentials (EPSPs.)

The most common CNS EPSP channels are glutamate receptors (gluR), which have an architecture (Fig. 2-6) reminiscent of an inverted K_{ir} molecule with extra TM segments. gluRs are heteromeric tetramers with three TMs per subunit. They have a large extracellular region with four glutamate binding sites and a cytoplasmic-facing P loop. Several functionally different gluRs are discussed in more detail in Chap. 6. They are all cation-selective; some allow Ca entry and others do not.

Less is known about the architecture of ATP-sensitive channels except it clearly differs from that of nAChRs and gluRs. The P2XR has two TMs per subunit, but the number of subunits per channel is not known. "P" refers to the purine sensitivity; adenine is a purine. P2 distinguishes them from P1 receptors, which are sensitive to adenosine and act through adenylyl cyclase. The P1 receptors are often referred to as

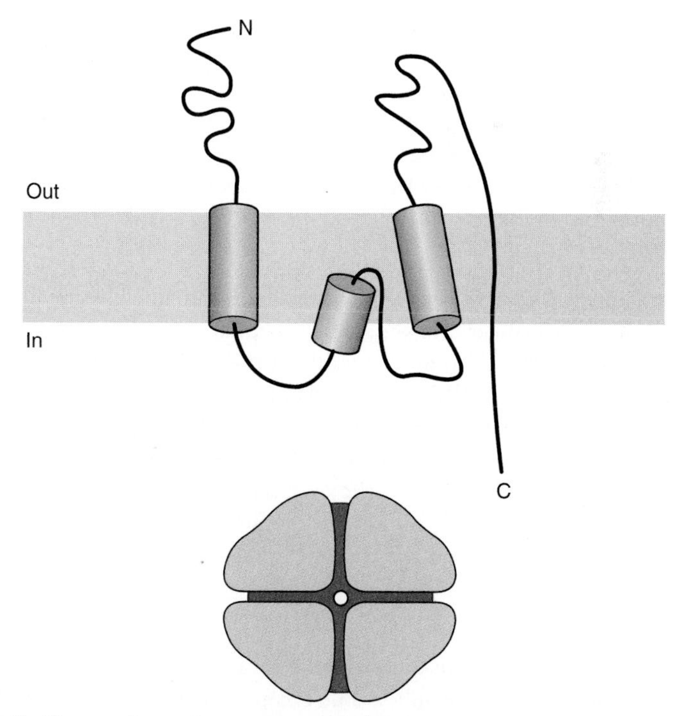

Figure 2-6. The topology of one monomer of glutamate receptor channels (gluR), with a top view showing the arrangement of the four monomers.

A receptors (A for adenosine); they are GPCRs. Caffeine is an antagonist of some of the A receptors. P2 receptors prefer ADP or ATP to adenosine. P2XRs are channels and P2YRs are GPCRs. **Purinergic receptors** are best known as regulators of blood flow in tissues; they have also been implicated in several sensory processes.

Two additional channel families have chemosensitive members but also have important members without known ligands. These are the **epithelial sodium channel** (ENaC) and the **IP_3 receptor** (IP_3R) family.

ENaCs are important in the reabsorption of sodium from the nascent urine in the tubules of the nephron. ENaCs are thought to be heteromeric tetramers each with two TM segments; they are not voltage-dependent. It is known that they are regulated by control of their insertion and removal from the membrane, and some people suspect that there is an unknown ligand for this channel. There are structurally related channels in invertebrates that have known ligands.

IP_3Rs and the related **ryanodine receptors (RyR)** are found in the membrane of the endoplasmic reticulum. When open, they permit the release of Ca from the endoplasmic reticulum. **Inositol triphosphate (IP_3)** is a second messenger produced by the action of phospholipase C (PLC) on the membrane lipid phosphotidylinistol, which has been previously phosphorylated to be PIP2. RyRs also control the release of calcium, primarily in muscle, from the sarcoplasmic reticulum. Ryanodine refers to a toxin that partially opens these channels. RyRs are opened by direct interaction with a modified Ca_V channel in skeletal muscle and by intracellular Ca in cardiac muscle. The function of IP_3Rs and RyRs are discussed in further detail in Chap. 7.

RyRs are homotetramers with an enormous 20-nm-diameter cytoplasmic N-terminal region. The total molecular weight for the tetramer is above 2 million, about 10 times larger than Na_V or K_V channels. IP_3R channels are also homotetramers about half the size of RyRs. It has been predicted that IP_3Rs have six TM segments per monomer and RyRs have four to twelve.

Water Channels

Some cells require more permeability to water than is provided by the lipid bilayer. Red blood cells, which must quickly change shape to pass through narrow capillaries, and some epithelial cells, most notably in the kidney, have specialized water channels or **aquaporins (AQPs)**, which permit the passage of water but exclude ions. The AQPs are found as tetramers with four functional pores, one in each subunit. The subunits have six TM segments and two regions similar to the P loop of K_V channels. One of the loops faces the extracellular surface, the other faces the cytoplasm, and they meet in the middle of the membrane. The function of AQPs and ENaCs are discussed toward the end of this chapter.

Cell-Cell Channels

9

In most tissues there are channels that connect the cytoplasm of one cell to the cytoplasm of its neighbor. The exceptions are free-floating cells in the blood and skeletal muscle cells. These channels are mostly between cells of the same type, but there are some cells of different type with junctions between them. These channels were originally detected electrically by showing that current could pass from one cell to another through an electrical synapse. Later they were associated with an anatomic structure called the gap junction, named for its appearance in electron micrographs. Actually this gap is spanned by matching arrays of proteins from each cell, with up to thousands of cell-cell channels per gap junction.

Each cell-cell channel is made of two **hemichannels**, one from each cell (Fig. 2-7). They are also called **connexons**. A hemichannel is a homomeric or heteromeric hexamer of proteins called **connexins**. There are more than 15 different connexins with molecular weights between 25 and 50 K. They all have four TM segments and two extracellular loops and their N and C terminals are in the cytoplasm. Some but not all connexins can form hybrid channels joining different hemichannels on the two cells.

The pore is much larger than the ion channels described above. It is about 1.2 nm and is permeable to anions, cations, and small metabolites as well as second

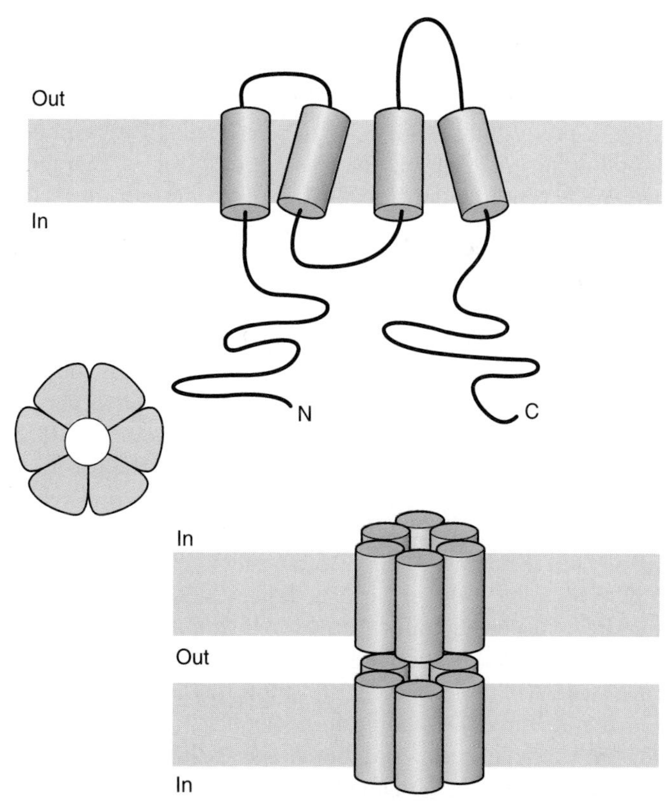

Figure 2-7. The topology of connexin, a monomer of cell-cell channels, top view showing the arrangement of six monomers in a hemichannel and side view showing two cell membranes with aligned hemichannels.

messengers such as ATP, cAMP, or IP_3 but not proteins. Experimentally, the pore is permeable to molecules with molecular weights below 1000. Cell-cell channels allow cells in a tissue to work in a coordinated manner.

If a cell is damaged, it can close its cell-cell channels leading to its neighbors and thus prevent the loss of small molecules from the whole tissue. This gating is controlled by intracellular Ca^{2+}, H^+, or transjunctional voltage. Different connexons have relatively different sensitivity to these three changes. Gating can also be induced by octanol and anesthetics such as halothane.

PUMPS

Ions move across cell membranes via channels, pumps, and transporters. These are three fundamentally distinct mechanisms and the student should be careful not to confuse them. Channels allow ions to move down their electrochemical gradient. Pumps create and maintain these gradients, moving ions up the gradient at the expense of ATP. Channels use these gradients to produce the various electrical signals. Transporters also use one or more gradients; the down-gradient movement of an ion (often Na) is coupled to the up-gradient movement of another substance. Because they consume ATP, the pumps are often referred to as ATPases.

Four pumps will be described in more detail: the Na/K pump, the Ca pump, and two types of proton pump. Three of these are called P-type pumps, because they are autophosphorylated during the reaction cycle, or E1-E2 pumps, because they have two major conformational states. The other proton pump is called F-type after the coupling factors F0 and F1, required for photosynthesis.

Na/K Pump

The Na/K pump, often referred to more simply as the Na pump, moves three Na ions out of the cell and two K ions into the cell in a cycle that converts one ATP molecule to ADP + Pi. At maximum speed, the pump completes about 100 cycles per second (cps), which means the movement of ions per molecule is much less than a Na_V channel, which may allow 1000 ions/ms to flow into the cell. The Na_V channels are open only briefly when the cell is active; the pump runs continuously to recover from the activity. Pump activity increases when intracellular Na or extracellular K increases and the pump acts homeostatically to restore the original levels.

The Na pump is a heterodimer with an alpha subunit that has the Na, K, and ATP binding sites and a beta subunit thought to be important for membrane insertion. The beta subunit has one TM segment; the alpha subunit probably has ten. Intracellular Na and ATP bind to the E1 form of the alpha subunit, which is then phosphorylated and converts to the E2 form (Fig. 2-8). The E2 form releases the Na into the extracellular space and binds extracellular K. The phosphate is hydrolyzed off the protein; the protein changes back to the E1 form, releases the K inside the cell, and the cycle continues. As the Na and K alternately move through the membrane, the pump passes through an occluded state where the ions are not accessible to either solution.

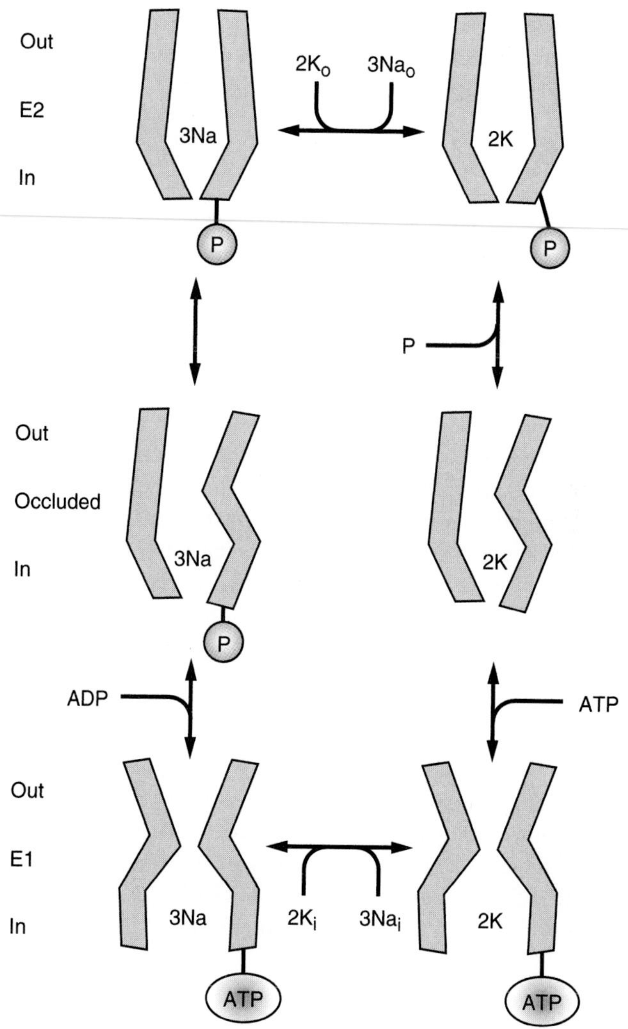

Figure 2-8. The Na/K pump cycle.

Digitalis and **ouabain**, a related **cardiac glycoside**, stop the action of the pump by binding extracellularly to the E2 form. Digitalis is used to treat a variety of cardiac conditions. It is a relatively dangerous drug and must be used cautiously so as to block only some of the pump molecules and leave others functional. The danger is complicated because extracellular K antagonizes the binding of digitalis by driving the pump toward the E1 form; the prudent physician will monitor blood K levels during digitalis treatment.

The Na pump is electrogenic, because each cycle moves one net charge out of the cell. This current has only a small effect on the membrane potential compared to ion flow through channels, which is discussed in the next chapter. The net movement of Na out of the cell prevents NaCl from accumulating in the cell. If the pump is blocked with cardiac glycosides, the cell will swell because of the osmotic influx of water following the NaCl.

Ca Pump

There are two important Ca pumps, one that pumps Ca from the cytoplasm into the extracellular space and another, the **SERCA pump**, that pumps Ca from the cytoplasm into the lumen of the sarcoplasmic or endoplasmic reticulum. They are thought to have similar mechanisms; both are P-type E1–E2 pumps that move two Ca ions out of the cytoplasm and two or three H ions into the cytoplasm for each ATP consumed.

The SERCA pump structure has been solved in several different states. It is a tall molecule, about 15 nm high and 8 nm thick, mostly extending out of the membrane on the cytoplasmic side. There are 10 TM segments. The cytoplasmic headpiece consists of the A (actuator), N (nucleotide binding), and P (phosphorylation) domains. The three cytoplasmic domains are widely split in the E1 • 2Ca state but gather to form a compact headpiece in the other states. This motion is transmitted to the membrane portion through helices 1 to 3, attached to the A domain, and 4 and 5, attached to the P domain, to allow the Ca to be released on the noncytoplasmic side. The distance between the Ca binding sites and the phosphorylation site is greater than 5 nm.

H/K Pump

The H/K pump secretes acid into the stomach by pumping two H ions out of the parietal cells of the gastric glands and two K ions into the cell while splitting one ATP molecule. Similar pumps also operate in epithelial cells in the intestine and kidney. This is an E1–E2 P-type pump and has a beta subunit, similar to the Na/K pump. The H/K pump is inhibited by omeprazole (Prilosec), the first FDA-approved over-the-counter treatment for frequent heartburn.

F-Type H Pumps

The most significant F-type H pump usually runs in reverse as the F0–F1 ATP synthase found in mitochondria and chloroplasts. This protein complex allows protons to flow down their electrochemical gradient and convert the flow of 10 protons to form 3 ATPs from ADP. The hydrogen gradients are produced by oxidative metabolism in mitochondria and by primary photosynthesis in chloroplasts.

The pump has eight different subunits and more than 20 polypeptide chains. The F0 portion spans the membrane and carries the H ions; the F1 extends into the mitochondrial matrix. Part of the complex rotates about an axis perpendicular to the plane of the membrane, similar to a turbine, as the H ions flow through. Another portion, the stator, stays fixed in position, and the interaction between the rotator and the stator produces a sequence of conformational states that favor the synthesis of ATP. In the presence of high ATP, low ADP, and no proton gradient, the process can be reversed to pump H.

A similar pump, the **V-type H pump**, moves protons into vacuoles and also into other intracellular organelles such as lysosomes, the Golgi apparatus, and secretory vesicles.

TRANSPORTERS

Transporters move ions and other small molecules across the membrane and are not channels or pumps. Sometimes the word *transporter* is used in the general sense to include all transport mechanisms and *secondary transporter* is used to distinguish this group. Transporters undergo a conformational change as they transport; in this aspect they are similar to pumps and different from an open channel. Unlike a pump, they do not consume ATP. Most transporters are thought to have 12 TM segments in two groups of 6 with a larger cytoplasmic loop between them. Some have a twofold pseudosymmetry and P loops facing both surfaces. There are three general categories of transporters: uniporters, symporters or cotransporters, and antiporters or exchangers (Fig. 2-9).

The **glucose transporter** (GLUT) is a uniporter that facilitates the diffusion of glucose down its concentration gradient into many cells that are consuming glucose but also out of cells that are releasing glucose by breaking down glycogen and out of basal surfaces of epithelial cells that line the intestines and kidney tubules (see Fig. 2-14).

The **Na-glucose cotransporter** (SGLT) is a symport that carries glucose into intestinal and kidney epithelial cells across their apical surfaces up the glucose concentration gradient. The energy required for this transport comes from the movement of one or two sodium ions down their electrochemical gradient for each transported glucose molecule.

The structure of a bacterial glutamate transporter, thought to be similar to the Na/glutamate cotransporter that recovers glutamate at CNS synapses, has recently been solved. It has eight TM segments with P loops between TM 6 and TM 7 facing the cytoplasm and those between TM 7 and TM 8 facing the outside. The crystals were made in the presence of glutamate, and a conspicuous nonprotein electron density, possibly glutamate but not clearly resolved, was seen at the interface between these loops. It is thought that relatively small movements of the protein can transfer the glutamate from one loop to the other and thus across the membrane.

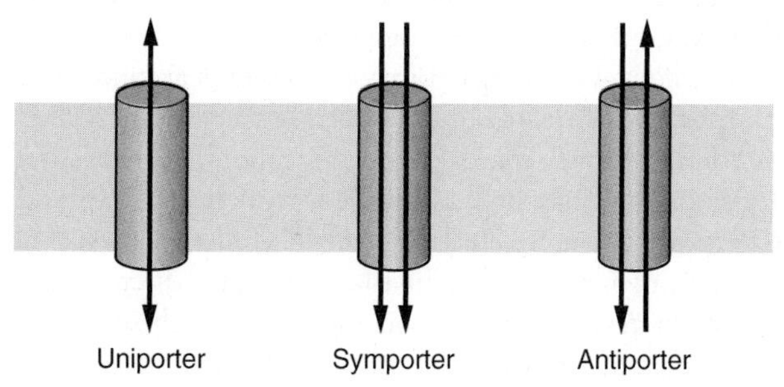

Figure 2-9. Three types of transporters.

The glutamate transporter in nerves couples the downhill motion of two Na ions and one K ion to the uphill transport of one glutamate.

There is an H/glutamate antiporter that uses the H gradient, established by a V-type pump, across the membrane enclosing synaptic vesicles to concentrate glutamate inside the vesicle.

There are many other Na-driven cotransporters to move other small molecules into cells and H-driven transporters to move some things into vesicles. Some of these transporters are targets for pharmacological intervention. For example, fluoxetine (Prozac) acts on an Na/serotonin cotransporter. Others are discussed further in Chap. 6. Some anions are cotransported with sodium; for example, the Na/I symporter concentrates iodine into thyroid follicle cells.

The **Na/Ca exchanger** (NCX) is an important regulator of intracellular Ca concentration. Three sodium ions moving down their electrochemical gradient into the cell can move one calcium ion out, or vice versa; all of the exchangers can run either way depending on the relative gradients. The effect of digitalis on cardiac muscle is to raise intracellular Na by inhibiting the Na/K pump. Elevated Na_i means that there is less inward gradient for Na and therefore less Ca efflux via NCX and thus an increased Ca_i and a stronger contraction (see also Chap. 7).

The Cl/HCO_3 exchanger, also known as the anion exchanger (AE), is important for moving CO_2 from the tissues to the lungs. CO_2, produced by metabolism in the cells, is converted to bicarbonate by carbonic anhydrase in the red blood cells, and the HCO_3 moves into the serum exchanging for chloride via AE. The process is reversed as the blood passes through the lungs and the CO_2 moves into the air to be exhaled.

ABC Transporters

This mixed group of 12 TM transport proteins all contains a characteristic ATP binding cassette amino acid sequence and, in the absence of more specific information, is assumed to consume ATP while transporting some material across the membrane. Two ABC transporters deserve mention here, the **multidrug resistance** (MDR) transporter, which is a pump, and the **cystic fibrosis transmembrane regulator** (CFTR), which is a channel.

MDR1 extrudes hydrophobic drugs across the cell membrane. It is thought to act somewhat like the flippase and extrudes the drugs without much specificity. A wide variety of cells in the GI tract, liver, and kidney express MDR proteins. These can frustrate the physician who is attempting to provide drugs to treat cancer among these cells.

CFTR is a protein which, when mutated, leads to cystic fibrosis. The wild-type protein is a chloride channel that requires phosphorylation by protein kinase A (PKA) and additional ATP hydrolysis by the activated CFTR protein in order to open. The Cl moves down its electrochemical gradient. Cystic fibrosis occurs because of the lack of Cl transport in the pancreatic duct (hence cystic). The decreased Cl leads to decreased water and the protein-rich secretion thickens and can block the ducts that then become fibrotic. Before the development of oral replacement therapy for the missing pancreatic enzymes, many CF patients died of complications of malnutrition. Now the major problem is the thickening of mucus in the lungs because of insufficient fluid secretion.

MEMBRANE RECEPTORS

The word *receptor* comes from pharmacologic studies, where it designates the site of action or the molecule that a small molecule of interest, perhaps a hormone or neurotransmitter, acts on. Here it is used in a more restrictive sense to mean molecules that span the membrane, are acted on the external surface by the small molecule, and trigger some action inside the cell when the small molecule is present. There are also intracellular receptors; for example, the steroid hormone receptor. Steroid hormones and related drugs can cross the lipid bilayer and bind these intracellular proteins. Chemosensitive channels are excluded as well, although some pharmacologists like to call them ionotropic receptors. There are two major categories of these membrane receptors: the G protein–coupled receptors (GPCRs) and the enzyme-linked or catalytic receptors.

G Protein–Coupled Receptors

(12) GPCRs have seven TM segments with an extracellular N terminus. They are coupled to a trimeric GTP-binding protein complex. When a hormone or neurotransmitter interacts with a GPCR, it induces a conformation in the receptor that activates a heterotrimeric G protein on the inner membrane surface of the cell (Fig. 2-10). In the inactive heterotrimeric state,

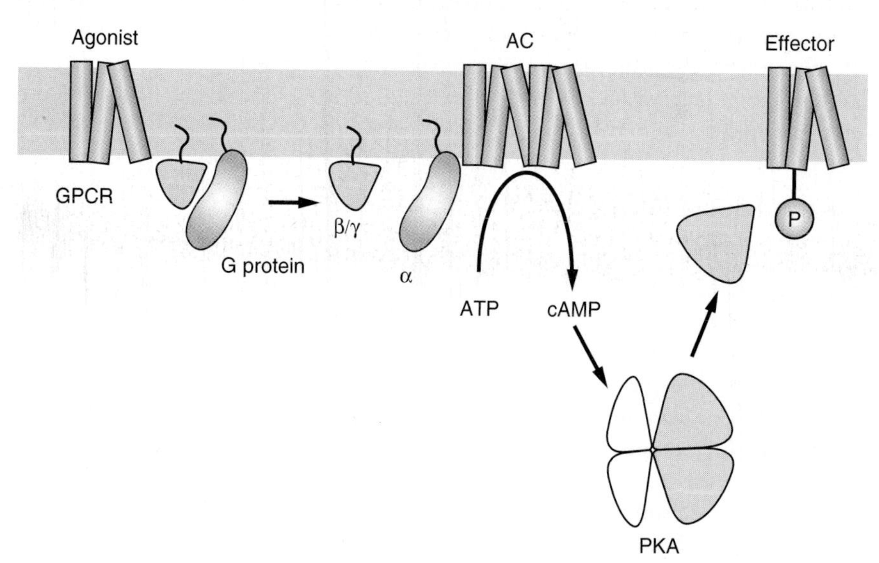

Figure 2-10. The Gαs signaling pathway. Binding of agonist to the G-protein–coupled receptor causes the dissociation of the α subunit, which causes adenylyl cyclase to raise cAMP levels. This, in turn, causes protein kinase A to phosphorylate an effector protein (in this case a channel).

GDP is bound to the Gα subunit. Upon activation, GDP is released, GTP binds to Gα, and subsequently Gα-GTP dissociates from G$\beta\gamma$ and from the receptor. Both Gα-GTP and G$\beta\gamma$ are then free to activate other membrane proteins. Most Gα and Gγ are lipidated; they have a covalently attached lipid anchor into the membrane bilayer. The duration of the G-protein signal is determined by the intrinsic GTP hydrolysis rate of the Gα subunit and the subsequent reassociation of Gα-GDP with G$\beta\gamma$.

There are more than 2000 predicted GPCRs in the human genome, more than 5 percent of all the genes. More than 800 are olfactory receptors; others detect almost all neurotransmitters and many hormones. Light also is detected by GPCRs in the eye. Different cells have different palettes of GPCRs coupled to different G proteins controlling different sets of intracellular reactions.

There are only about 16 Gα subunits and fewer G$\beta\gamma$. Three classes of Gα subunits initiate most of the subsequent events described in this book. Gα_s stimulates **adenylyl cyclase** (AC), Gα_i inhibits AC and its associated $\beta\gamma$ directly activates K_{Ach} channels, and Gα_q stimulates a **phospholipase** (PLCβ). AC produces cAMP that can directly influence some channels. cAMP also activates **phosphokinase A** (PKA), which phosphorylates many proteins, thus altering the activity of the cells. PLCβ splits the membrane phospholipid phosphatidylinositol to produce IP_3 and diacylglycerol (DAG). As described above, IP_3 binds the IP_3R channels, which increases Ca_i, which in turn triggers various reactions. Several examples of GPCR-initiated cascades are described more fully in Chaps. 4, 6, and 7.

The toxins that underlie two infectious diseases, **cholera** and **pertussis**, ADP-ribosylate Gα subunits leading to constitutive activation. In cholera, activated Gα in intestinal epithelial tissue stimulates AC, cAMP levels increase, and CFTR chloride channels open, leading to a watery diarrhea. People with cystic fibrosis are resistant to cholera because they have fewer functional chloride channels. The cellular pathogenesis of pertussis is not clear.

Enzyme-Linked Receptors

Most enzyme-linked receptors are **receptor tyrosine kinases** (RTK) and act by phosphorylating tyrosine side chains on other proteins, which may in turn phosphorylate other proteins. Some enzyme-linked receptors are not kinases themselves but are coupled to an associated protein that phosphorylates other proteins. Some enzyme-linked receptors are guanylyl cyclases, tyrosine phosphatases, or serine kinases. Most growth and differentiation factors act by binding specific RTKs.

The **insulin receptor** is an RTK that phosphorylates a family of substrates known as insulin-receptor substrates; these stimulate changes in glucose, protein, and lipid metabolism and also trigger the Ras signaling pathway, activating transcription factors that promote growth.

The **CD4** and **CD8** molecules on the surface of T lymphocytes are examples of receptors that are coupled to a cytoplasmic tyrosine kinase. CD stands for clusters of differentiation that refers to the technique of using fluorescent antibodies to differentiate functionally different lymphocytes one from another. CD4 and CD8 would be better named as major histocompatibility complex (MHC) enzyme-linked receptors.

Cell Adhesion Molecules

Most cells except red blood cells have integral membrane proteins that attach to the extracellular matrix or with adhesion molecules on neighboring cells. These molecules hold the tissue together and can allow the transmission of mechanical forces from one cell to another. They can act as signals during development, so one cell can recognize another. Many also act as receptors, informing the inside of the cell that they have bound something. Some are controlled from the inside, binding only when some signal has been received.

The **integrins** are examples of cell-matrix adhesion molecules. They have a single TM segment and link cells to fibronectin or laminin in the extracellular matrix.

Cadherins are Ca-dependent cell-cell adhesion molecules; they are glycoproteins with a single TM segment and are thought to bind homophilically to cadherins on the other cell. Cadherins have been found at many neuron-neuron synapses. There is a large family of cell adhesion molecules, of which the **N-CAMs** are the best-studied. N-CAMs are found on a variety of cell types and most nerve cells. Like cadherins, N-CAMs have a single TM segment and bind homophilically, but they differ in that they do not require Ca for binding. **Intercellular adhesion molecules** (ICAMs) are a related class expressed on the surface of capillary endothelial cells that have been activated by an infection in the surrounding tissue. ICAMs bind heterophilically to integrins on white blood cells and help them move to the site of infection. **Selectins** are carbohydrate-binding proteins on the endothelial cell membrane that recognize sugars on the surface of the white blood cell and form the initial binding, which is strengthened by the ICAMs.

TRANSPORT ACROSS CELL MEMBRANES

From a functional point of view, discussion of the transport of materials across cell membranes can be divided into passive transport, where the materials move down their concentration gradient, and active transport, which creates or maintains these gradients.

Passive Transport

SIMPLE DIFFUSION

13 Some materials can move down their concentration gradient by simple diffusion though the lipid bilayer. Small uncharged molecules such as O_2, CO_2, NH_3, NO, H_2O, steroids, and lipophilic drugs can enter or leave cells by simple diffusion. The net flux of these compounds through the membrane is proportional to difference in their concentration on the two sides, or, as expressed in an equation:

$$J_{1\text{->}2} = -P\,(C_2 - C_1) = -P\,\Delta C \qquad [2.1]$$

Using the centimeter-gram-second (CGS) system of units, $J_{1\text{->}2}$ is the number of moles that move through a square centimeter of membrane from side 1 into side 2 each second and C_1 and C_2 are the numbers of moles of the material per cubic centimeter of solution on the two sides. P, the proportionality constant, is called the **permeability** of the membrane to this material in centimeters per second. The equation is written with the leading minus sign as an aid to remembering that the flux is moving down the concentration gradient. This relationship is shown graphically in Fig. 2-11.

Equation [2.1] is **Fick's first law**. It can be used to describe the flux by simple uncharged substances through any membrane. For example, it is useful to describe the movement of oxygen from the air into the alveoli of the lungs and into the blood, across the cells of the alveolar epithelium and the capillary endothelium. A charged species will also be influenced by the electrical potential difference across the membrane in a manner to be discussed in the next chapter. If there is zero potential difference across the membrane, Fick's law is applicable also to charged substances.

Permeability describes a property of a particular membrane in relation to a particular substance. The membrane is considered *permeable*, while the substances are said to be *permeant* or *to permeate*. The permeability will be proportional to the ability of the substance to partition into the membrane and to diffuse within the membrane. The permeability will be inversely proportional to the thickness of the membrane. It is usually not easy or necessary to know these three factors separately, but one should appreciate that thickening of the complex membrane between the alveolar cavity and the blood will reduce the movement of oxygen into blood.

It is sometimes convenient to think of Fick's law as saying that the **net influx** of a substance is equal to the **unidirectional influx** (PC_o) minus **the unidirectional efflux** (PC_i). Permeabilities are often measured using radioactive tracers and arranging the experiment so that the tracer concentration on one side is kept near zero and the unidirectional flux is measured directly.

The permeability is a measure of the ease with which a solute crosses through a membrane. Plain lipid bilayers are relatively permeable to small uncharged molecules;

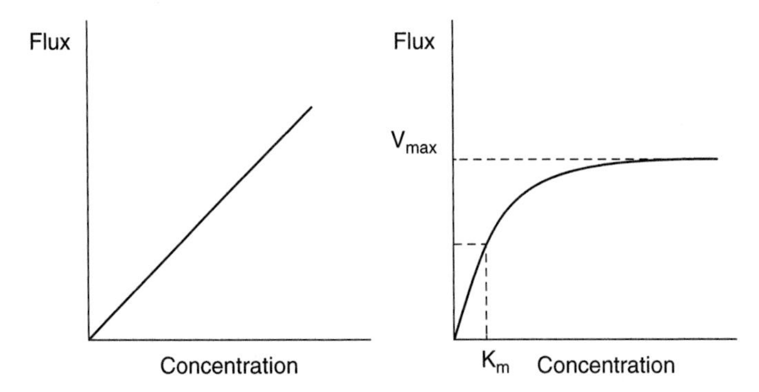

Figure 2-11. The concentration dependence of simple diffusion (left) and facilitated diffusion (right).

the permeability to water is about 10^{-3} cm/s. Thus water equilibrates across a cell membrane in a few seconds. Urea is moderately permeable, $P = 10^{-6}$ cm/s, and its equilibration time is a few minutes. Hydrophilic small organic molecules such as glucose or uncharged amino acids are less permeable, with $P = 10^{-7}$ and equilibration times of hours; ions are essentially impermeable, with $P = 10^{-12}$ cm/s and equilibration times of many years.

Facilitated Diffusion

(14) Many substances, such as glucose or urea, easily enter cells in spite of having low oil–water partition coefficients; therefore the lipid bilayer is relatively impermeable to them. The flux of these materials is described by Fick's law only for low concentrations. At higher concentrations, the flux saturates at a maximum value (see Fig. 2-11). This behavior can be described by the Michaelis-Menten equation, which is also used to describe enzyme kinetics. The unidirectional flux is given by the following equation:

$$J = J_{max}\, C/(C + K_m) \qquad [2.2]$$

J_{max} is the maximum flux and K_m is the affinity or the concentration at which the flux is half its maximum value. This saturable property of the flux suggests that there are a fixed number of sites at which the flux can take place. Also, as in the case of enzymes, it may be possible to demonstrate the competition of different substances for the same site or noncompetitive inhibition of the transport sites. The sites are selective for a particular substance or group of substances that they will transport or that allow competition for transport. Selectivity, affinity, and J_{max} are three independent qualities of the sites; they will be found with different values in different systems. Facilitated diffusion is now understood in terms of channels or transporters.

Most channels have low affinity or high K_m values; they are not saturated under normal physiological conditions. Three glucose uniporters, GLUT1, GLUT3, and GLUT4, are found in nearly every tissue and have a high affinity for glucose; they are saturated at all physiological concentrations. GLUT2, which is found in tissues carrying large glucose fluxes (such as intestine, kidney, and liver), has a low affinity for glucose, and the influx through GLUT2 transporters increases as the blood glucose concentration increases.

Active Transport

Pumps provide **primary active transport**, moving materials up their electrochemical gradients at the expense of ATP. Cotransporters and exchangers can provide **secondary active transport**, using a gradient produced by primary active transport to move another material up its gradient. The flux by pumps and transporters can be described by equations similar to Eq. [2.2], modified to include the affinity for each substance and for ATP. When more than one ion at a time is involved in the reactions at one pump or transporter molecule, the equation must also be modified to reflect this cooperativity. Thus the efflux of sodium through the Na/K pump has a sigmoidal relationship to the internal Na concentration.

Water Transport

Life is intimately associated with the movement of water. Our bodies are mostly water and are vitally dependent on its continual supply. Water is a small but abundant molecule. It is not much larger than an oxygen atom, about 0.2 nm across—small enough to intercalate between other molecules, even in some crystals. A mole of water is 18 mL; thus pure water is 55 mol/L. Because the molecules are small, they move easily. Because they are so abundant, their movements are important to our well-being. The death of infants by dehydration subsequent to diarrhea is a health problem in many parts of the world. It can be alleviated with clean water and a little sugar and salt. Edema (swelling), often seen in sports injuries but also in more severe conditions, is due to the movement of water.

There are three distinct mechanisms for water movement: bulk flow, molecular diffusion, and molecular pumping. When you pull the plug in a bathtub or your heart beats, there is bulk flow of water in response to an external mechanical force—a push or a pull that is capable of stretching a spring. The driving force for bulk flow is the mechanical pressure commonly produced by pushing or by gravity.

Molecular diffusion or **osmosis** is a passive process by which water diffuses from areas of high water concentration to those of low. There is a high water concentration where there is low solute concentration, and vice versa. Water can diffuse across most cell membranes directly through the lipid bilayer or by traveling through specific water channels or **aquaporins**. Many cells produce aquaporins, because simple diffusion does not permit adequate water flux. Some kidney cells will insert aquaporins in response to **antidiuretic hormone** (ADH), so as to increase water flow from the forming urine back into the blood, thereby conserving water. This passive type of water movement is called osmotic flow, and the associated driving force is the concentration gradient of the water.

Water can also be transported across membranes at the expense of energy by the **Na-glucose cotransporter** (SGLT1). The transmembrane transport of two Na ions and one sugar molecule is coupled, within the protein itself, to the influx of 210 water molecules, independent of the osmotic gradient. The energy comes from allowing Na to move down its concentration gradient. This molecular pumping is a secondary active transport mechanism and could account for almost half the daily uptake of water from the small intestine.

15 Osmotic pressure is the mechanical pressure that produces a flow of water equal and opposite to the osmotic flow produced by a concentration gradient. This concept is similar to (and historically preceded) the Nernst equilibrium potential, an electrical potential that produces a flow of ions equal and opposite to a flow produced by a concentration gradient. The Nernst potential is discussed further in Chap. 3. If two different solutions are in contact, the osmotic pressure, π, between them is

$$\pi = RT\sum \sigma_n \Delta c_n \qquad [2.3]$$

R is the molar gas constant (Avogadro's number times Boltzmann's constant), T is the absolute temperature, Δc_n is the concentration difference of the nth solute, and σ_n is the **reflection coefficient** of the membrane for that solute. The reflection coefficient indicates the impermeability of the solute compared to water, ranging from 0.0 for a solute as permeant as water to 1.0 for a solute that is completely impermeant. In spite of all the channels discussed above, the permeability of electrolytes is very low compared to that of water and their reflection coefficient is approximately 1.0. For a simple case with only impermeant molecules, Eq. [2.3] reduces to

$$\pi = RT\ \Delta c$$

The concentration refers to the summated molar concentration of all the particles created when the solute is dissolved in water. It is measured as the **osmolarity**, that is, the sum of the moles of each component of the solution. A 2-mM solution of $MgCl_2$ contains 6 **milliosmoles** (mosm) per liter, 2 for the Mg^{2+} and two for each Cl^-. The osmolarity of this solution is 6 **milliosmolar**. A 3-mM NaCl solution and a 6 mM urea solution have the same osmolarity and are said to be isosmotic.

The **osmolality** of a solution can be measured by the change it produces in the freezing point or vapor pressure. Osmolality refers to moles of solute per kilogram of solvent, whereas osmolarity refers to moles of solute per liter of solution. Clinically the distinction is moot, and you may hear the terms used interchangeably. Also the actual pressures are rarely discussed; rather, the osmoles are mentioned directly.

16 **Tonicity** is a concept that is related to osmolarity but is a special case for cells. A solution is said to be **isotonic** if it causes neither shrinking nor swelling of cells. A 150-mM NaCl solution (9 g/L or 0.9%) is isotonic for mammalian cells and also **isosmotic** to the cell contents. A 300-mM urea solution is also isosmotic to the cell contents, but a cell placed in this solution will swell and eventually **lyse** or burst (Fig. 2-12). The urea solution is **hypotonic**; it has insufficient tonicity to keep the cell from swelling. It differs from the NaCl

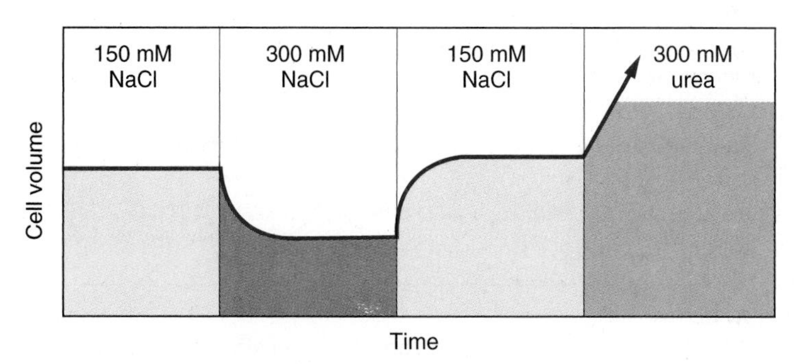

Figure 2-12. Cells shrink in hypertonic solutions and swell in hypotonic solutions.

solution because the urea can cross the cell membrane. The addition of a permeable material to a solution increases its osmolarity but not its tonicity. The addition of more impermeable solutes makes a hypertonic solution; a 300-mM NaCl solution is hypertonic and will cause cells to shrink.

If a moderately permeable solute is added to an isotonic solution (e.g., 300 mM urea + 150 mM NaCl), the cells will transiently shrink and then return to their original volume (Fig. 2-13). The rate at which they shrink is proportional to the water permeability of the membrane; the rate at which the volume recovers is proportional to the urea permeability.

In some cases it is convenient to consider a reflection coefficient as a description of the permeability of solutes. Water movement across capillary walls depends on the mechanical or hydrostatic pressure difference and on the difference in colloid osmotic pressure due to differences in protein concentration in the plasma and the interstitial fluid. If the capillary wall is completely impermeable to the proteins, it is said to have a reflection coefficient of 1.0. If the walls become leaky, the reflection coefficient decreases, proteins enter the interstitial space, and water follows.

Water movement in the whole body is concerned with two compartments, intracellular and extracellular. The extracellular compartment has two subcompartments: the plasma fluid in the blood vessels and the **interstitial fluid** that bathes the rest of the cells. The plasma and the interstitial fluid are separated by the capillary walls, which are freely permeable to all the small molecules and ions but normally prevent the plasma proteins from entering the interstitial fluid. The proteins have an overall net negative charge at blood pH. The equilibrium that arises with an impermeable protein and freely permeable ions is called **the Gibbs-Donnan equilibrium.** This effect produces small concentration gradients (<3 percent) and a small potential (a few mV, lumen negative) across the capillary walls. For most clinical purposes, this can be ignored and the ionic concentrations in the plasma, which are easily measured, can be considered to represent the extracellular fluid in general.

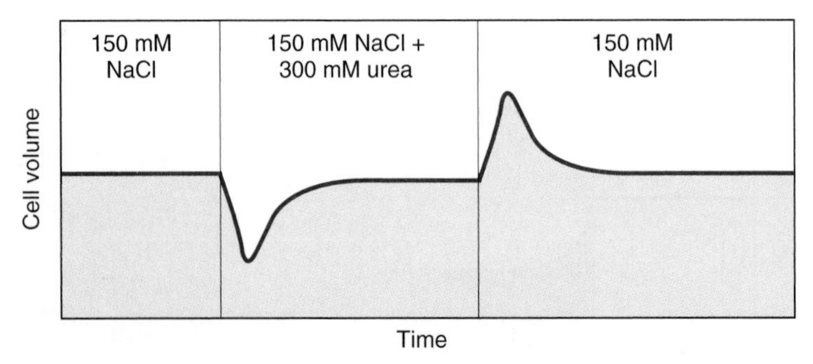

Figure 2-13. The addition of urea causes transient shrinking but does not change the steady-state tonicity.

The plasma proteins are osmotically important; they tend to keep water in the blood vessels. The balance between the hydrostatic pressure and this **"colloid" osmotic pressure**, called the **Starling effect**, regulates the flow of water across capillary endothelia. **Edema** is the loss of this balance.

The volumes and osmolarity of the body compartments can be calculated using the following four principles: In every compartment the volume times the osmolarity is the number of osmoles. Water will move between the compartments to make the osmolarity equal in all compartments. The total amount of water and the total number of osmoles is the sum of the amounts in the compartments. Any added osmolytes will remain extracellular; added water will distribute among the compartments according to the first three principles.

For example, consider a 70-kg medical student with 16 L of extracellular fluid and 24 L of intracellular fluid for a total of 40 L. If the osmolarity of these compartments is 300 mosM, the extracellular compartment contains $16 \times 0.3 = 4.8$ osm and the intracellular compartment contains 7.2 osm, for a total of 12 osm. If this student should happen to swallow 1 L of seawater containing 1 osm of salts, mostly NaCl, the total water increases to 41 L and the total number of osmoles increases to 13, so the new osmolarity is $13/41 = 317$ mosM. The extracellular compartment will have 5.8 osm, so its new volume will be $5.8/0.317 = 18.3$ L. The new intracellular volume will be $7.2/0.317 = 22.7$ L. Notice that water moved out of the cells to dilute the seawater.

The healthy body has homeostatic mechanisms to restore the original balance. The kidney will excrete a more concentrated urine and the student will experience thirst and probably drink water without osmolytes.

Hyponatremia, or low sodium in the blood, can lead to serious problems in marathon runners, who lose water and salt due to perspiration but tend to replace only the water. This leads to hypotonic blood, so that water leaves the blood vessels and swells the cells. If this happens in the brain, which is in an enclosed cavity, unconsciousness and death may ensue.

TRANSPORT ACROSS EPITHELIAL CELLS

17 Many epithelial cell layers form functional membranes between two solutions and act in a coordinated way to selectively transport solutes and water across the layer. This is achieved by having **tight junctions** between the sides of the epithelial cells so that the sheet of cells is impermeable to substances that cannot pass through the cell membranes and by incorporating selective pumps and channels appropriately on the two surfaces of the sheet. The two sides may be called by different names in different epithelia. **The apical membrane** faces the lumen or outside of the body; it can be known as the **luminal** or **mucosal** membrane or the **brush border** after the appearance of its microvilli. The **basolateral** membrane that faces the inside of the body can be known as the **serosal** or **peritubular** membrane.

Figure 2-14 shows pathways for Na and glucose transport across epithelial cell layers. Na/K pumps in the basolateral membrane keep the internal Na low by

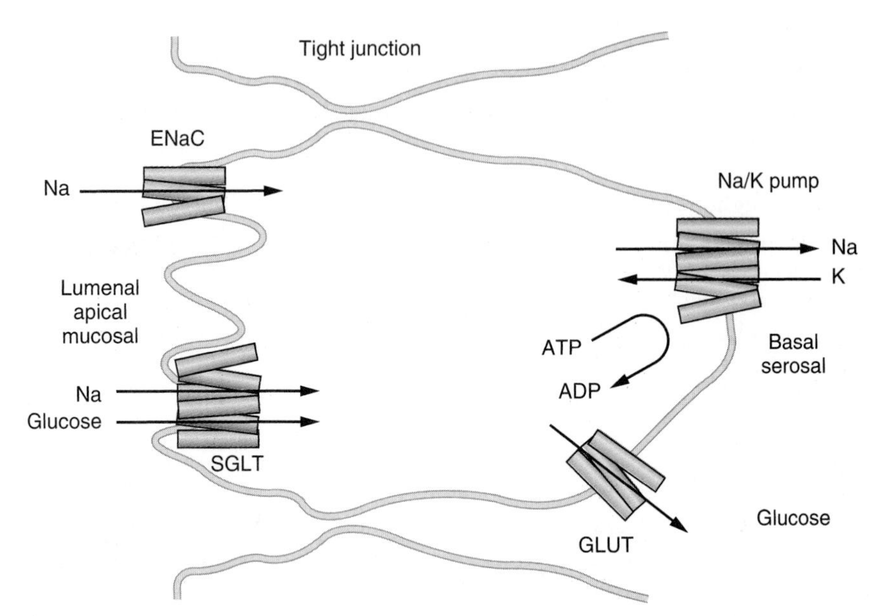

Figure 2-14. Sodium and glucose are transported through epithelial cell layers by a combination of pumps, channels, and transporters.

moving it into the extracellular fluid. Na can enter the cell by moving down its concentration gradient through ENaC channels on the apical membrane and leave via the pump on the other side. Glucose may be brought into the cell through the apical membrane, up its concentration gradient by the sodium/glucose cotransporter (SGLT), and then move down its concentration gradient through the glucose uniporter (GLUT) on the basolateral surface.

When solutes are moved across epithelial membranes, water may flow osmotically, "following" the solute. This effect is important in the kidney, where water retention is achieved by solute retention. It is also important for rehydration therapy to combat the water loss of diarrhea. Adding glucose and salt to the drinking water will stimulate SGLT to move Na, glucose, and water into the cell. The Na/K pump and GLUT transporter will then move the solutes into the body and the water will follow.

Water and water-soluble material may move across epithelia by **transcytosis** or by **receptor-mediated endocytosis**. The substance is taken up into vesicles by **endocytosis** on one surface and either released unchanged by **exocytosis** on the other surface or broken down in endosomes and the products released by transporters. Transcytosis occurs across capillary endothelia; receptor-mediated endocytosis is important in the kidney and liver.

KEY CONCEPTS

A lipid bilayer surface membrane with embedded proteins separates and connects cells with the surrounding extracellular environment.

The lipid molecules are amphipathic, with hydrophobic groups facing the interior of the membrane and hydrophilic groups facing both aqueous interfaces.

The proteins carry out specific functions by acting as channels, pumps, transporters, receptors, or cell adhesion molecules.

The proteins are amphipathic, generally with one or more hydrophobic transmembrane helices.

Ion channels are membrane proteins with a pore that selects for the type of ion(s) that passes through the channel down their electrochemical gradient.

Mechanosensitive channels are generally cation-selective and have diverse structures.

Voltage-sensitive channels have a fourfold symmetry. Each of the parts has six transmembrane helices, one of which carries multiple positively charged amino acids.

There are many different families of chemosensitive or ligand-gated channels corresponding to the different chemical ligands.

Cell-cell channels connect the interior of one cell to the interior of an adjacent cell by an aqueous path that permits the passage of ions and other small molecules.

Pumps move ions or other molecules up their gradients at the expense of ATP.

Transporters move some ions or other molecules up their gradients at the expense of having other ions move down their gradients.

G-protein–coupled receptors initiate intracellular G-protein cascades under the control of extracellular activators.

Some materials can simply diffuse down their concentration gradients through the membrane lipids.

Many substances have specific facilitated diffusion mechanisms characterized by an affinity and a maximum transport velocity.

Osmotic pressure is proportional to the osmolarity or the total concentration of all solutes.

Tonicity describes a solution's ability to prevent the shrinking or swelling of cells.

Substances may be transported through epithelial cell layers by combinations of pumps and transporters arranged on opposite sides of the cells.

STUDY QUESTIONS

2–1. What are the properties that distinguish free diffusion, facilitated diffusion, primary active transport, and secondary active transport?

2–2. How do ion channels achieve selectivity among the various ions?

2–3. Calculate the changes in intracellular and extracellular compartments for the student described in the text when she drinks 4 L of distilled water, starting from the resting condition. Repeat for drinking 1 L of a 300-mosM sports beverage.

2–4. What is the general function of gap-junction channels? What is the function of gap-junction gating?

2–5. Why are there so many different connexins in an organism?

SUGGESTED READINGS

Boyd D, Schierle C, Beckwith J. How many membrane proteins are there? *Protein Sci* 1998;7:201–205.

Burnstock G. The past, present and future of purine nucleotides as signalling molecules. *Neuropharmacology* 1997;36:1127–1139.

Cabrera-Vera TM, Vanhauwe J, Thomas TO, et al. Insights into G protein structure, function, and regulation. *Endocr Rev* 2003;24:765–781.

Chang AB, Lin R, Keith Studley W, et al. Phylogeny as a guide to structure and function of membrane transport proteins. *Mol Membr Biol* 2004;21:171–181.

Fahy E, Subramaniam S, Brown HA, et al. A comprehensive classification system for lipids. *J Lipid Res* 2005;46:839–861.

Farfel Z, Bourne HR, Iiri T. The expanding spectrum of G protein diseases. N Engl J Med 1999;340:1012–1020.

Golub T, Wacha S, Caroni P. Spatial and temporal control of signaling through lipid rafts. *Curr Opin Neurobiol* 2004;14:542–550.

Kellenberger S, Schild L. Epithelial sodium channel/degenerin family of ion channels: A variety of functions for a shared structure. *Physiol Rev* 2002;82:735–767.

Simons K, Vaz WL. Model systems, lipid rafts, and cell membranes. *Annu Rev Biophys Biomol Struct* 2004;33:269–295.

Sobczak I, Lolkema JS. Structural and mechanistic diversity of secondary transporters. *Curr Opin Microbiol* 2005;8:161–167.

Soderlund T, Alakoskela JI, Pakkanen AL, Kinnunen PKJ. Comparison of the effects of surface tension and osmotic pressure on the interfacial hydration of a fluid phospholipid bilayer *Biophys J* 2003;85:2333–2341.

Sukharev S, Corey DP. Mechanosensitive channels: Multiplicity of families and gating paradigms. *Sci STKE* 2004;2004(219):re4.

Toyoshima C, Inesi G. Structural basis of ion pumping by Ca^{2+}-ATPase of the sarcoplasmic reticulum. *Annu Rev Biochem* 2004;73:269–292.

Unwin N. The Croonian lecture 2000. Nicotinic acetylcholine receptor and the structural basis of fast synaptic transmission. *Philos Trans R Soc Lond B Biol Sci* 2000;355:1813–1829.

Vial C, Roberts JA, Evans RJ. Molecular properties of ATP-gated P2X receptor ion channels. *Trends Pharmacol Sci.* 2004;25:487–493.

Yamagata M, Sanes JR, Weiner JA. Synaptic adhesion molecules. *Curr Opin Cell Biol* 2003;15:621–632.

Channels and the Control of Membrane Potential

3

OBJECTIVES

- *Describe how membrane potentials are measured and provide typical values for different cells.*
- *Discuss the relationship between the separation of charge across the membrane and the membrane potential.*
- *List the approximate concentrations of the major ions in the intra- and extracellular compartments.*
- *Describe the three factors that control the movement of ions through membranes.*
- *Determine whether an ion will move into or out of cells given the membrane potential and the concentration gradient of the ion.*
- *Discuss how the membrane potential changes when ions flow across cell membranes.*
- *Explain the steps that occur during the generation of a Nernst potential.*
- *Explain the steps that occur during the generation of a resting membrane potential.*
- *Discuss why the net flux of charge is 0 in the resting state even though ions are moving through the membrane.*
- *Discuss the role of the Na/K pump in the generation of the membrane potential.*
- *Define single-channel recording and describe currents through single K channels.*
- *Describe the two types of the spread of electrical information in nerve and muscle cells.*
- *Discuss why the cell membrane acts as a capacitor and what properties this conveys on nerve and muscle cells.*
- *Discuss the difference between length (space) and time constants and the relationship of these constants to nerve conduction.*
- *Explain the steady state and transient cable properties of nerve and muscle cells.*

All living cells have an electrical potential difference across their surface membranes. Cells act as miniature batteries; the battery cell is named after the biological cell. At rest the inside of cells is negative to the outside by about 0.01 to 0.1 V or 10 to 100 mV. Concentration gradients of ions across the membrane are the immediate supplier of the energy to create and maintain the resting potential. The resting potential is necessary for electrical excitability of nerve and muscle cells, sensory reception, CNS computation, and to help regulate transfer of ions across the membrane.

MEASURING MEMBRANE POTENTIALS

Figure 3-1 shows how resting potentials are measured. A muscle is secured to the bottom of a dish that is filled with an isotonic salt solution with an ionic composition similar to that of blood. A **microelectrode** with a fine tip pulled out of glass and filled with 3 M KCl is positioned over one of the muscle cells. A chlorided silver wire in the microelectrode is attached to one terminal of a voltage-measuring device, in this case an oscilloscope that displays a trace of voltage vs. time. The other terminal is attached to another chlorided silver wire placed in the dish; this is called the ground wire. When the microelectrode is in the solution, it is at the same potential as the ground wire and the oscilloscope reads 0 mV. When the microelectrode is advanced a few micrometers into the muscle cell. the trace on the oscilloscope abruptly jumps to about –90 mV and stays there as long as the microelectrode remains in place. When the electrode is withdrawn, the trace returns to 0 mV. The experiment can be repeated. If a second microelectrode is inserted, it measures the same potential, showing that the electrodes are not somehow creating the potential.

When the microelectrode is inside the cell, the KCl is in contact with the cytoplasm that is in contact with the membrane. The ground wire is in contact with the external solution, which is in contact with the outside of the membrane. The potential difference is across the membrane; it is called the membrane potential. The particular membrane potential measured when the cell is at rest—that is, not active—is also called the resting potential. Different cells have different resting potentials. Skeletal and cardiac muscle cells have a resting potential of about –90 mV. Sensory and motor neurons have a resting potential of about –70 mV; smooth muscle cells, about –60 mV; and red blood cells, about –10 mV.

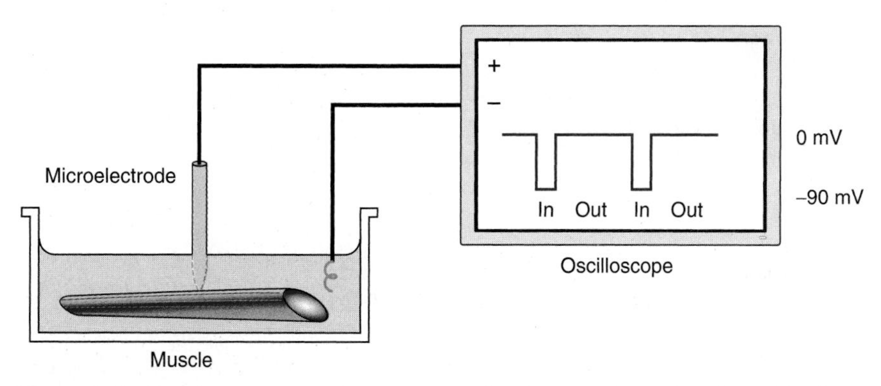

Figure 3-1. Membrane potentials are measured with microelectrodes filled with electrolyte solutions.

SEPARATION OF CHARGE

The resting membrane potential is a reflection of the separation of charges across the membrane. There are a few excess negative charges (about 1 pmole/cm^2) on the inner surface and the same number of excess positive charges on the outer surface (Fig. 3-2). The solutions on the two sides contain about 150 mmol/L of cations and anions (Table 3-1) with exactly balanced positive and negative charges except for the layer within about 1 nm from the surface of the membrane. The bulk solutions on both sides are electrically neutral.

The excess charges of opposite sign experience an attractive force for each other but are prevented from reaching each other because they cannot easily leave the aqueous solutions and enter the oily lipid membrane. Any charge within the membrane also experiences this force, tending to pull positive charges inward and push negative charges outward. The voltage across the membrane is the electrical measurement of this **electromotive force** or this **potential** for movement of charges if they happen to be within the membrane.

The voltage is directly proportional to the amount of charge that is separated. The ratio of separated charge to the voltage is called the membrane **capacitance.**

$$C = Q/V \qquad [3.1]$$

Electrical charge is measured in terms of the **coulomb** (C); there are 96,484 C/mol of charge (this is **Faraday's constant**). The unit of capacitance is the farad (F); 1 C/V is 1 F. Capacitance is the ability to store separated charges. Many small computers use a capacitor to store enough charge to allow some minimal function to remain for a short time while the battery is being changed. The membrane stores the opposite charges by keeping them separated.

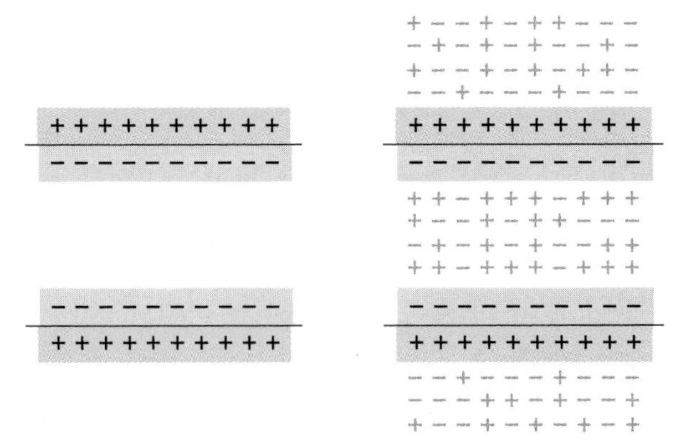

Figure 3-2. The separation of charge. Left, a single layer of charges separated by the membrane. Right, adding a representation of the mobile charges in the bulk solutions.

Table 3–1. Concentrations of Some Ions of Importance across a Muscle Cell Membrane[a]

Ion	Extracellular concentration	Intracellular concentration	E_{ion}
Cations			
Na^+	145 mM	12 mM	+65 mV
K^+	4.5 mM	155 mM	−95 mV
Ca^{2+}	2.5 mM	100 nM	+132 mV
Anions			
Cl^-	132 mM	4 mM	−90 mV
A^-	~0 mM	155 mM	
HCO_3^-	22 mM	8 mM	−26 mV

[a] A^- represents impermeant anions inside the cell. Many are polyvalent; all together, they do not contribute 155 mosm to the osmotic pressure. There are also other uncharged osmolytes in the cell.

GENERATION OF THE RESTING POTENTIAL

The membrane separates two solutions with quite different ionic compositions. The generation of the resting potential and all of the changes in potential (such as the action potential and the synaptic potentials) depend on the concentration gradients of ions across the cell membrane. Table 3-1 presents some typical values for a skeletal muscle. Both sides are electrically neutral, the sums of positive and negative charges are equal. The external solution has a relatively high Na and Cl concentration and a modest K concentration, whereas the internal solution is high in K and low in Na and Cl; it has a high concentration of other anions (A^-), such as phosphate groups on proteins or nucleic acids and negatively charged amino acids on proteins.

There is an inward concentration gradient for Na and Cl and an outward concentration gradient for K. The sodium gradient is about 10-fold; Cl, about 30-fold; and K, about 40-fold. Table 3-1 indicates a 25,000-fold inward concentration gradient for Ca. Exact numbers are given in this table to facilitate the calculation of examples later in this chapter. There is a normal variation of about 10 percent in different people or different muscles for Na, K, and Cl values. External Ca is normally about 2.5 mM, but internal Ca can change dramatically with activity, increasing above 1 μM when the muscle is contracting.

The cell membrane is permeable to all of the ions listed in Table 3-1 except A^-; they can move through the membrane via various channels. The membrane is not very permeable to ions; compared with water, their permeability is insignificant. However, it is the control of this ionic permeability that regulates the membrane potential and small (by chemical standards) movements of ions that change the membrane potential.

FACTORS THAT CONTROL ION MOVEMENTS

The movement of ions is proportional to the net **driving force** upon them. The net driving force is the **electrochemical gradient** or the difference between the driving force due to the concentration gradient and the force due to the voltage gradient or membrane potential. The movement of charged particles is an electrical current, I. The relationship between the current carried by a particular ion, x, and the driving force can be expressed as

$$I_x = g_x\,(V - E_x) \qquad [3.2]$$

E_x is the chemical driving force for ion x expressed as an electrical potential; this is described more fully below. V is the membrane potential and $(V - E_x)$ is the driving force on ion x. The **membrane conductance** for ion x is g_x. The overall membrane conductance for ion x is proportional to the number of channels for that ion, N; the probability that a channel is open, P_o; and the conductance of a single open channel, γ; or

$$g_x = NP_o\gamma \qquad [3.3]$$

The conductance is proportional to the permeability of the membrane or the ease with which ions move through it. The conductance is also proportional to the concentration of the conducting ion(s). In the absence of sodium ions, a sodium channel may be permeable (if it is open), but it will not conduct any current.

The voltage gradient pushes or pulls an ion because the ion is charged. The concentration gradient is a conjugate force; ions tend to move from a high concentration to a low concentration. More ions will hit an open channel from the side with higher concentration than the side with lower concentration, so there will be a flow down the concentration gradient in proportion to the gradient.

To determine the net flux of an ion through the membrane it is necessary to know the concentration gradient, the voltage gradient (the membrane potential), and the conductance for the ion. Unless all three factors are known, it is not possible to predict the flux of the ion. The two forces on the ion from the voltage and concentration gradients may act in the same direction or in opposite directions.

THE NERNST EQUILIBRIUM POTENTIAL

For any particular concentration gradient it is possible to pick a voltage gradient that is equal and opposite, so that the term in parentheses in Eq. [3.2] is zero and there is no net current. This is called the **electrochemical equilibrium potential** or the **Nernst potential** and is given by

$$E_x = \frac{RT}{Fz}\ln_e\frac{C_o}{C_i} \qquad [3.4]$$

E_x is the Nernst potential (or the **equilibrium potential** or the diffusion potential) for the ion, C_o and C_i are the concentrations on the outside and inside of the cell, z is the charge of the ion or the valence, R is the molar gas constant, T is the absolute temperature, and F is Faraday's constant. RT is the thermal energy of the material at temperature T and RT/F is this energy expressed in electrical units. At room temperature, RT/F is about 25 mV. The equation can be simplified to

$$E_x = \frac{60 \text{ mV}}{z} \log_{10} \frac{C_o}{C_i} \qquad [3.5]$$

with

$$z = +1 \text{ for } Na^+ \text{ or } K^+, +2 \text{ for } Ca^{2+}, -1 \text{ for } Cl^-$$

and so on.

The equilibrium potential for an ion is the potential at which the net flux is zero. It can be calculated theoretically using the formula of Eq. [3.5] without knowledge of the actual membrane potential. It is a way to express the concentration gradient in electrical terms, so that the concentration gradient can be compared to the voltage gradient.

The Nernst potentials for the various ions in Table 3-1 are listed in the last column. Figure 3-3 compares three of these equilibrium potentials with a resting potential of –90 mV.

For chloride, the concentration gradient is inward; Cl ions would like to move into the cell because there is a higher concentration outside. The –90 mV resting potential exerts an outward force on the negatively charged chloride ions. These two are equal and opposite; that is, $(V - E_{Cl}) = -90 - (-90) = 0$ mV, and chloride ions are in electrochemical equilibrium.

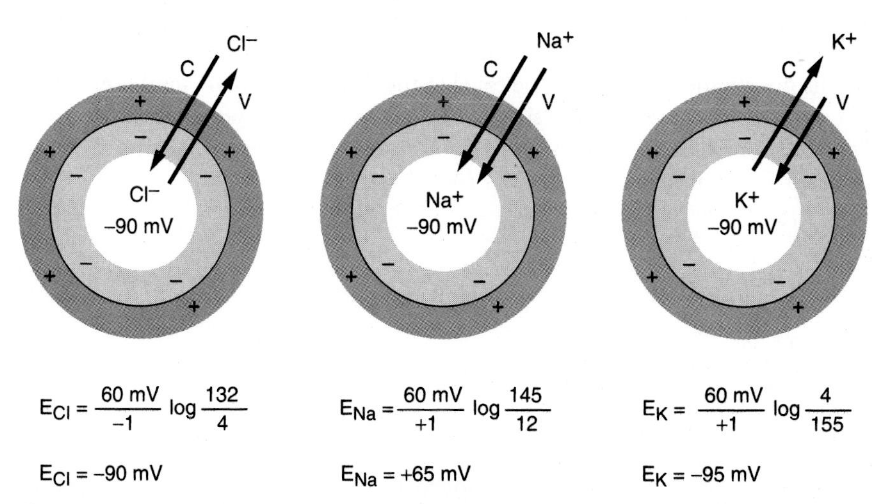

Figure 3-3. The driving force on ions crossing through the membrane, voltage gradients, and concentration gradients.

For sodium, the concentration gradient is also inward, but the negative membrane potential exerts an inward force on the positively charged Na ion. Both forces are inward and sodium ions are far from equilibrium; that is, $(V - E_{Na}) = -90 - (+65) = -155$ mV. If the membrane were permeable to Na, it would readily enter.

For potassium, the concentration gradient is outward while the force from the voltage gradient is inward. The magnitude of the concentration gradient is slightly larger than that of the voltage gradient; that is, $(V - E_K) = -90 - (-95) = +5$ mV. Potassium ions are not at equilibrium; they have a tendency to leave the cell.

Chloride is the only ion in Table 3-1 that is at equilibrium. Cl ions are distributed at or very near equilibrium in skeletal muscle cells but not in most nerve cells.

Generation of the Nernst Potential

The resting potential has its particular value because of the K and Na gradients and because the resting membrane is much more permeable to K than to Na. This is more easily understood by first considering a membrane separating the same gradient that is permeable only to K ions. Such a membrane could be constructed by reconstituting biological K channels into an artificial lipid bilayer or by using a potassium ionophore, such as valinomycin, to make a lipid membrane permeable to K (Fig. 3-4).

When the solutions are first added to the compartments, there is zero membrane potential. K^+ will start to move down its concentration gradient and thereby move positive charge from compartment B to compartment A, leaving an excess negative charge on side B and producing an excess positive charge on side A. This separation of charge means that there is now a membrane potential with side B negative to side A (or, equivalently, side A positive with respect to side B). As side B becomes more negative, the further net flow of K from B to A will be reduced, until eventually sufficient charge will have been separated so that the flux due to the increasing electrical attraction is equal and opposite to the flux due to the concentration gradient. At this point electrochemical equilibrium will have been reached and the membrane potential will be equal to the Nernst potential—in this example, −95 mV with side B negative to side A. This could also be expressed by saying that the Nernst potential is +95 mV, with side A positive to side B. This is a property of Eq. [3.5], because $\log A/B = -\log B/A$.

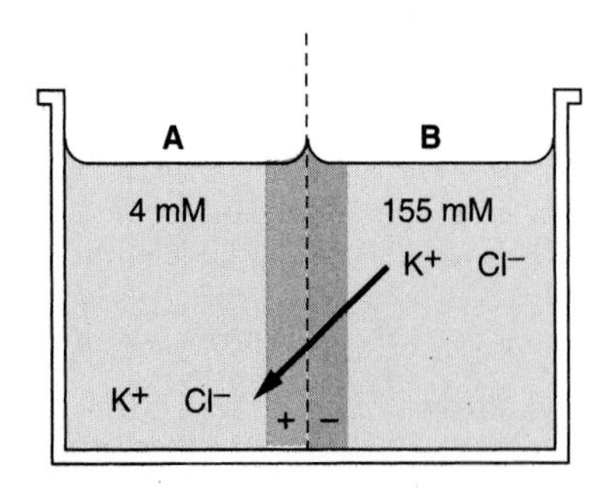

Figure 3-4. K flowing down its concentration gradient through an artificial bilayer that is permeable only to K^+ ions.

Notice that less than 1 pmol/cm^2 flux of positive charge is required to establish the membrane potential. The bulk concentrations of K on the two sides of the membrane have not changed significantly. The change in concentrations is undetectable by ordinary chemical experiments.

An arrangement similar to that of Fig. 3-4 is used clinically to measure the concentration of K in blood or other solutions. Side B is prepared with a known concentration and the unknown is placed on side A. The system is allowed to come to equilibrium and the potential between the two compartments is read. Equation [3.4] is used to solve for the unknown K concentration. The pH meter is a similar arrangement, using a membrane, usually a special glass, that is selectively permeable to H^+ ions. Other electrodes are available for other ions as well.

THE RESTING POTENTIAL

The example above can be extended to explain the resting potential in a muscle cell by considering the situation that would occur if the membrane potential were artificially held at zero electronically and then released. Such a condition can be arranged with the voltage-clamp apparatus described Chap. 5. In order to understand the process, it is necessary to know the concentration gradients listed in Table 3-1 and also that the permeability of the membrane to K is 50 to 100 times greater than its permeability to Na.

Starting at 0 mV membrane potential, K will start to move out of the cell while Na will start to move in, both moving down their concentration gradients. However, more K will move than Na because the permeability to K is much greater than the permeability to Na, so a net positive charge will move out of the cell, making the inside of the cell negative with respect to the outside.

The developing negative membrane potential opposes the further efflux of K ions and acts to increase the influx of Na ions. This trend will continue, with the membrane potential becoming more and more negative until 3 Na ions are entering through the Na channels for every 2 K ions that are leaving through the K channels. At this point a steady state will be reached because the Na/K pump is extruding 3 Na and taking up 2 K ions on each ATP-consuming cycle. There is no net flux in this **steady state**, so the membrane potential will not change as long as the ATP supply is adequate (Fig. 3-5).

It is important to realize that the major role of the pump is indirect; the pump is very important for maintaining the gradients but contributes only a few millivolts directly to the membrane potential. If this experiment were repeated with the pump blocked by **ouabain** (a cardiac glycoside similar to digitalis) or the absence of ATP, the initial processes would be the same and the process would continue until the influx of sodium were equal to the efflux of potassium. At this point the membrane potential would stop, becoming more negative, and then very slowly start to move back towards 0 mV as the concentrations on the two sides of the membrane changed over several hours.

Using the figures in Table 3-1, it is possible to estimate the immediate difference in membrane potential that can be attributed to the running pump. When the

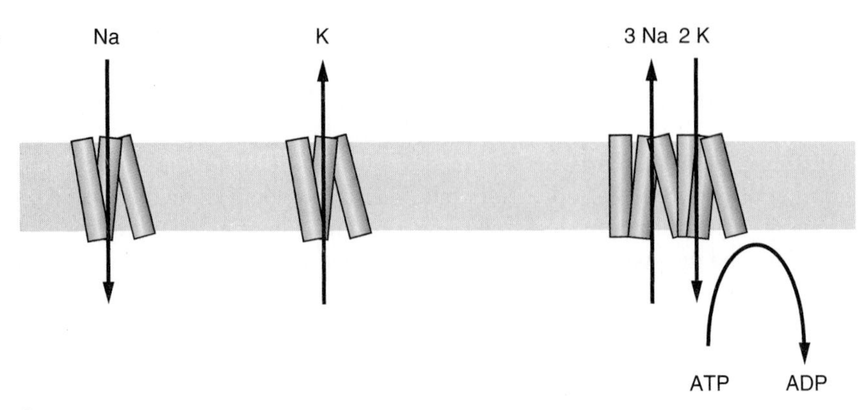

Figure 3-5. Ions flowing down their concentration gradient through channels and actively transported up their concentration gradient by pumps.

membrane potential is –90 mV, there is a 5-mV net driving force on K ions. If the membrane potential became 2.5 mV less negative to –87.5 mV, the driving force on K ions would be increased by 50 percent, so that three K ions would leave for every two that left at –90 mV. There would be a 2.5-mV decrease in the driving force on Na ions, but this is less than 2 percent of the 155-mV driving force, so it would make a negligible change in the Na influx, and three Na ions would enter for every three K ions leaving. Thus about 87.5 mV of the resting potential comes from the gradients and an additional 2.5 mV comes directly from the pump.

If the concentrations and the ionic conductances are known, the membrane potential can be calculated using Eq. [3.4] to find the Nernst potentials and Eq. [3.2] to find the currents. When the membrane potential is not changing, there is no net current. If the pump is not running and the membrane only conducts Na and K, $I_{Na} = -I_K$ or $g_{Na}(V - E_{Na}) = -g_K(V - E_K)$, which can be rearranged to solve for V.

$$V = (g_{Na}E_{Na} + g_KE_K)/(g_{Na} + g_K) \qquad [3.6]$$

The membrane potential is the weighted average of the equilibrium potentials, weighted by their respective conductances. If $g_K >> g_{Na}$ the membrane potential will be near E_K; if $g_{Na} >> g_K$ it will be near E_{Na}, and if they are equal, it will be halfway between. If the membrane is permeable only to these two ions and there is no external source of electrical current, the membrane potential will always be between E_K and E_{Na}. These concepts will become more useful when the conductances change, as seen in the next three chapters.

Because the resting membrane is preferentially permeable to potassium, the resting potential is sensitive to the external potassium concentration (Fig. 3-6). Increasing external K will bring the membrane potential closer to zero or **depolarize** the membrane. The resting membrane in its normal ionic environment is considered polarized. A change of potential in the positive direction, towards 0 mV, is a

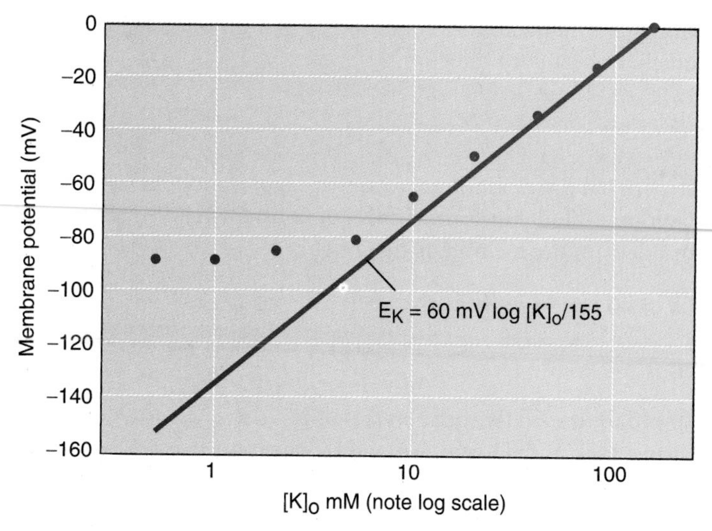

Figure 3-6. The membrane potential as a function of the external K concentration. Notice the logarithmic concentration scale.

depolarization. A change in the other direction, making the membrane potential more negative, is a **hyperpolarization**.

Elevated K_o depolarizes membranes because it reduces the K gradient across the membrane and makes E_K closer to zero. This reduces the tendency for K to leave the axon, so the balance is reached at a less negative potential. Elevated K_o is a dangerous, potentially lethal condition because excitable cells require the normal resting potential to remain excitable. Doubling the blood K level is likely to compromise cardiac function.

K_{ir} Channels Support the Resting Potential

Some cells, notably cardiac and skeletal muscle cells, have K_{ir} channels that are open, thus conducting, at the resting potential and are thought to be the major contributor to the resting K conductance. These were named inward rectifiers when experiments demonstrated that the inward current through them, when the membrane potential was hyperpolarized beyond E_K, was larger than the outward current seen when the membrane was depolarized. It is perhaps an unfortunate name because, in normal life, the membranes never experience such a large hyperpolarization. The important aspects of this channel's function are to be open for outward movement of potassium near the resting potential and then to become nonconducting when the cell is depolarized. This blocking in the depolarized state will be seen to be important for cardiac muscle action potentials, as described in Chap. 5.

K_{ir} is not a voltage-sensitive channel. The blocking comes about because Mg^{2+} or other polyvalent cations in the cytoplasm attempt to go through the channel when it is depolarized and get stuck, thus preventing K from using the channel.

If the channels are studied under conditions without polyvalent cations, they conduct K equally well in both directions.

GOLDMAN-HODGKIN-KATZ EQUATION

If permeabilities are known, rather than conductances, the theoretical Goldman-Hodgkin-Katz (GHK) or constant field equation is often used to calculate the membrane potential.

$$V = 60\text{ mV} \log_{10} \{(P_{Na}\,Na_o + P_K\,K_o + P_{Cl}\,Cl_i)/(P_{Na}\,Na_i + P_K\,K_i + P_{Cl}\,Cl_o)\} + \text{contribution due the pump} \qquad [3.7]$$

As in Eq. [3.6], the GHK equation simplifies to the Nernst equation if only one permeability is greater than zero. The GHK equation has been useful to describe experimental results when some of the concentrations are set to zero, which makes the Nernst potentials in Eq. [3.6] meaningless.

The relationship between permeability and conductance can be set on a quantitative basis by considering the condition when the membrane potential is zero and then, after multiplying the chemical flux by Faraday's constant, equating Eqs. [2.1] and [3.6] to obtain the electrical current. Thus

$$g_x E_x = P_x F\,\Delta C_x \qquad [3.8]$$

CHANGES IN MEMBRANE POTENTIAL

The membrane potential will change if current is injected into the cell, either through an experimenter's microelectrode or by opening channels that allow ions to flow down their electrochemical gradients. It takes time to change the membrane potential; it will not jump instantaneously to a new value. Many nerve and muscle cells are quite long, more than 1 m for some nerve cells. The effect of a localized current will spread passively from the site of injection but may not change the potential of the entire cell. These temporal and spatial effects are shared by electrical cables and are referred to as the **cable properties**. They can be understood by considering the membrane capacitance, the membrane resistance, and the longitudinal cytoplasmic resistance between different parts of the cell.

The **passive spread** by cable properties must be distinguished from the **active spread** by action potentials. The passive effects occur without any change in the number of open channels. If sufficient current is injected into a nerve axon to depolarize it above threshold, an action potential will be elicited and will propagate without loss of amplitude over the entire length of the cell. The action potential is **regenerated** as it propagates. As the wave of opening sodium channels moves, energy is supplied to the process from the Na gradient all along the axon. In contrast, a smaller depolarization or a hyperpolarization that does not open Na channels will spread only a few millimeters, becoming progressively smaller when measured at a greater distance from the stimulus.

The membrane capacitance is the ratio of the charge separated to the membrane potential—Eq. [3.1]. The capacitance is related to the membrane geometry by

$$C = \frac{K \text{ area}}{\text{thickness}} \qquad [3.9]$$

where K is a constant describing the material composition of the membrane. If the area is larger, it will take a greater amount of charge to change the potential. The thinner the membrane, the closer the charges are to each other and the more charges will have to be moved to change the potential. The capacitance of a typical membrane is about 1 $\mu F/cm^2$; this value is often used to estimate the size of a cell by measuring its capacitance.

The membrane resistance is the reciprocal of the membrane conductance.

$$R_m = 1/g_m \qquad [3.10]$$

The longitudinal resistance is proportional to the length and inversely proportional to the cross-sectional area.

$$R_l = \frac{\rho \text{ length}}{\text{area}} \qquad [3.11]$$

PASSIVE PROPERTIES OF A SMALL ROUND CELL

Consider first a cell that is small enough that all of its interior can be considered to be at the same potential. It will have a capacitance proportional to its area and non-infinite resistance due to the conductance of its open channels. Consider two microelectrodes inserted into the cell, one to inject current and the other to measure the membrane potential (Fig. 3-7*A*). The equivalent circuit, shown in Fig. 3-7*B*, can represent the electrical properties of this cell.

When a square pulse of current is injected into the cell, the voltage will change, as indicated in Fig. 3-7*C*. The magnitude of the current is the number of coulombs of charge per second. At first this flow of charge is all supplied to the capacitance and the voltage changes proportional to the amount of charge that has been injected. However, when the voltage changes, current will start to flow through the membrane resistance proportional to the change in voltage (Ohm's law). Eventually a new steady state will be reached where the charge on the capacitor and the membrane potential stop changing and all the subsequently arriving charge flows through the resistance. The steady-state potential differs from the original potential by $\Delta V = iR$.

When the pulse is shut off, the capacitance discharges through the resistance. The flow through the resistance decreases as the voltage decreases. This makes the decay exponential in time, or

$$\Delta V = iR \exp(-t/\tau)$$

The rising phase is the mirror image, or

$$\Delta V = iR[1 - \exp(-t/\tau)]$$

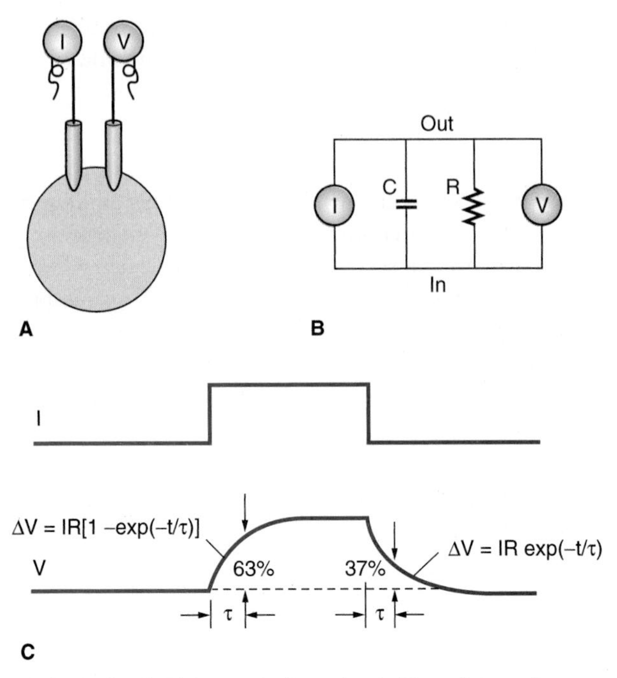

Figure 3-7. A spherical cell (*A*), its equivalent circuit (*B*), and the voltage response to an injected pulse of current (*C*).

τ is the characteristic **time constant**—the time it takes to discharge the change in voltage to $1/e = 37$ percent of its value (or the time it takes to charge to 63 percent of its final value). $\tau = RC$; a membrane with a larger resistance or a larger capacitance takes longer to charge or discharge. Many cells have time constants in the range of 1 to 20 ms.

PASSIVE PROPERTIES OF A LONG CYLINDRICAL CELL

In an extended cell, the response is a function of both time and distance from the site of stimulation. This is simpler to describe by first considering an artificial situation where time is not important. Figure 3-8*A* shows a long cell impaled by four microelectrodes, one to inject current and three to measure the membrane potential at different distances. When a constant current has been delivered long enough for it to reach a new steady state, the change in membrane potential is largest at the site of current injection and falls off exponentially in both directions.

$$\Delta V = \Delta V_o \exp\left(-\frac{x}{\lambda}\right)$$

λ is the characteristic **length constant**, the distance it takes for the potential to drop to 37 percent of its value at the site of injection. The potential will be 37 percent of this value or 14 percent of the original value two

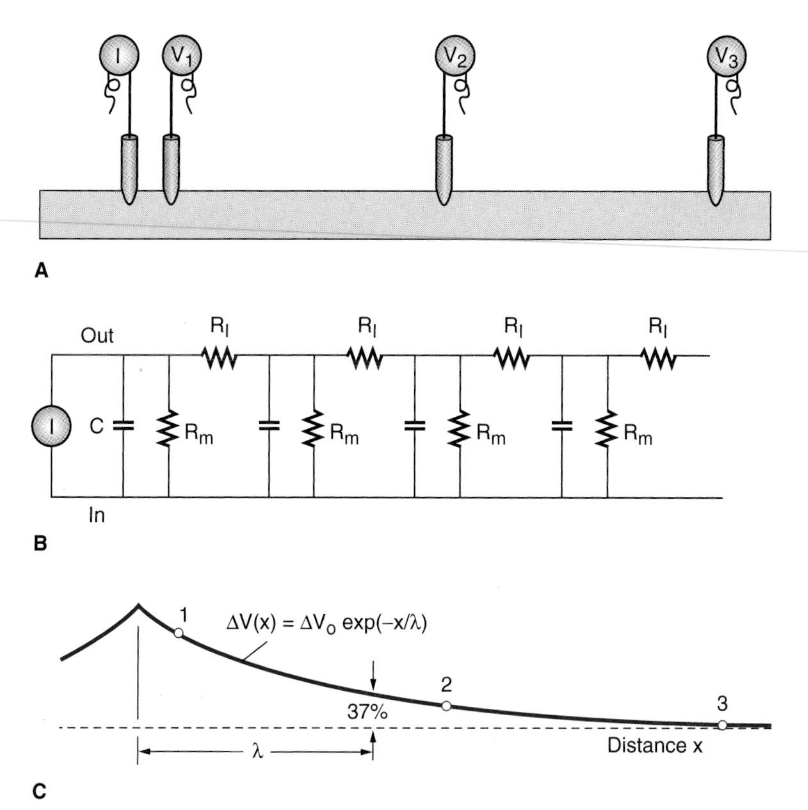

Figure 3-8. A long cell (*A*), its equivalent circuit (*B*), and the steady-state distribution of its membrane potential in response to a steady injection of current (*C*).

length constants away and less than 1 percent of the original value 5 length constants away. The length constants for most nerve and muscle cells are 0.1 to 2 mm. A 10-μm cell is approximately isopotential, but a 150-cm-long nerve cell requires an active propagation mechanism to be able to communicate electrical activity from end to end.

The voltage change declines because some of the injected current leaks out of the cell and is not available to depolarize the adjacent regions. The amount that leaks out is proportional to the voltage change, so the decline is exponential. The length constant depends on the ratio of the membrane resistance over the longitudinal axoplasmic resistance: $\lambda = \sqrt{(r_m / r_l)}$.*

If the membrane resistance is higher, the membrane will be less leaky, the length constant will be longer, and the potential will spread further. If the longitudinal

* The square root can be appreciated by considering that the units of r_l are ohms divided by centimeters and of r_m are ohms times cm.

resistance is lower, which is the case in larger-diameter axons, current will flow more easily down the axon and the length constant will be longer.

As the distance from the injection increases, the amplitude of the transient response decreases and the rise time becomes longer and more sigmoidal (Fig. 3-9). Initially most of the charge entering the cell goes to the membrane immediately adjacent to the source; only later is enough available to charge the distal membrane. When the pulse is terminated, all responses decay at the same rate. Synapses are distributed on the dendritic tree at different distances from the cell body. The more distant synapses will have less effect on the cell's activity, and it will be less abrupt in arriving.

Passive spread is also important for action potential propagation; it is the mechanism of connection between the active region and the adjacent resting region. Action potentials propagate more rapidly in larger-diameter axons because they have lower longitudinal resistance and longer length constants.

When cell-cell junctions join cells, they can operate electrically as if they were all one cell. Many of the cells in the heart are coupled and action potentials propagate from one cell to another supported by the passive spread of depolarization via the cell-cell junctions. There are also cell-cell junctions between some neurons in the CNS.

For some it is helpful to visualize a hydraulic analogy of these electrical phenomena. Electrical voltage is analogous to water pressure and electrical current to solution flow. The long cell is similar to a leaky hose, with lower membrane resistance corresponding to more leaks and lower longitudinal resistance corresponding to larger hose diameter.

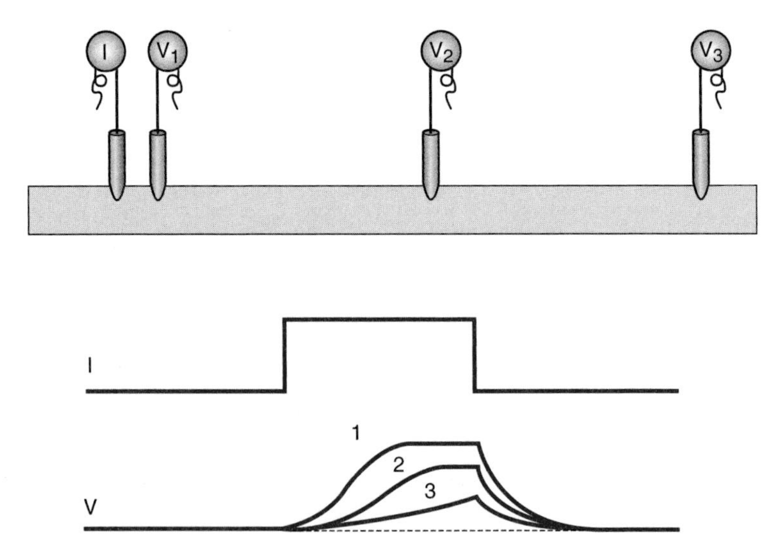

Figure 3-9. The transient voltage responses at three distances from the site of an injected pulse of current.

KEY CONCEPTS

An electrical membrane potential is directly proportional to the separation of positive and negative charges across the cell membrane. The ratio of separated charge to voltage is the membrane capacitance.

Cell membranes separate solutions with quite different ionic compositions.

The movement of ions is directly proportional to the net driving force on the ions. The net driving force is the electrochemical gradient or the difference between the effect of the membrane potential and the effect of chemical gradient.

The effect of the chemical gradient can be expressed by the Nernst equilibrium potential.

Only a very small number of ions must be separated to produce the membrane potential. This is negligible compared with the concentrations available on both sides.

The resting membrane potential is a steady state with ions moving down their electrochemical gradient through channels and an equal number being pumped up their electrochemical gradient at the expense of ATP.

The Goldman-Hodgkin-Katz equation can be used to calculate the membrane potential if the permeabilities to the various ions and their concentrations are known.

When current flows through the membrane, the membrane potential changes in time and in space, governed by the "cable properties."

When a step in current is injected into a cell, the time it takes the potential to achieve 63 percent of its final value is equal to the product of the membrane resistance times the membrane capacitance.

When a steady current is injected into a long cell, the potential change decays exponentially with distance, dropping by 63 percent in a length equal to the square root of the ratio of the membrane resistance divided by the longitudinal axoplasmic resistance.

STUDY QUESTIONS

3–1. What is the relationship between ion fluxes and electrical current?

3–2. How is the resting potential generated?

3–3. *What is the change in resting potential with each of the following changes in ion concentration? In each case there is also a change in the concentration of an impermeant counterion in order to maintain solution neutrality.*

a. 40-mM increase of Na^+_o

b. 10-mM increase of K^+_i

c. 10-mM increase of K^+_o

d. 100-mM increase of A^-_i

e. 50-mM decrease in Na^+_o

3–4. *What is the effect of ouabain poisoning on the resting potential?*

3–5. *For a set of ion concentrations (mM) as shown below,*

	In	out
Na	12	145
K	155	4.5
Cl	4	132
Anion	155	
Ca		2.5

Find the membrane potential for $g_K = 200\ g_{Na}$, $gNa = 50\ g_K$, $gNa = g_K$

3–6. *Draw axes similar to those below. On your axes, draw the changes in membrane potential for the sequence: $g_K >> g_{Na}$, $g_{Na} >> g_K$, $g_K >> >> g_{Na}$, $g_K >> g_{Na}$*

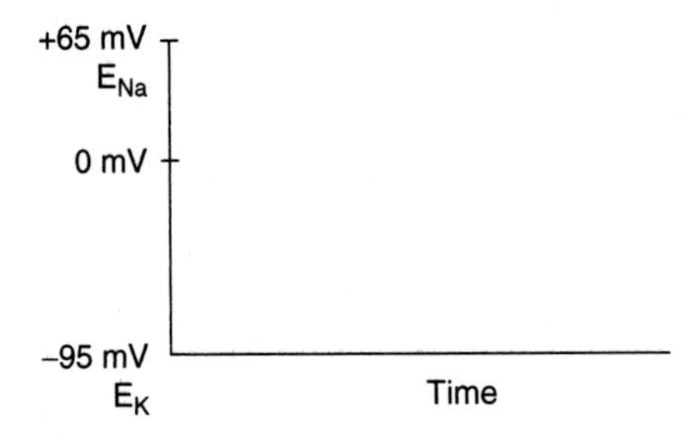

Sensory Generator Potentials 4

OBJECTIVES

- *List eight sensations and the names of the specialized sensory receptor cells responsible for generating these sensations.*
- *Describe sensory adaptation in these receptors.*
- *Draw a schematic cartoon of (a) a Pacinian corpuscle and its sensory ganglion cell (including cell body and central process); (b) a cochlear hair cell and its synapses; and (c) a photoreceptor and its synapses.*
- *List three or more differences between ion channels that underlie action potentials, resting potentials, and receptor potentials.*

Animals have developed a wide variety of sensory organs capable of monitoring chemicals, light, sound, and other mechanical events in the external and internal environment. The cells or portions of cells that perform the initial step of sensory transduction converts light or mechanical energy or the presence of specific chemical conditions into a change in the membrane potential called the **receptor potential** or **sensory generator potential**. In small sensory cells, this generator potential directly controls the synaptic release process described in Chap. 6. In longer cells, the generator potential will initiate an action potential that propagates to a distant presynaptic ending and then trigger the release process.

Each sensory cell has an appropriate stimulus, called its **adequate stimulus**. The CNS interprets signals coming from this cell in terms of its adequate stimulus. The adequate stimulus for photoreceptors in the eye is visible light. If an electric shock or sufficient pressure is applied to the eye, a person will report flashes of light, even if the room was dark.

Each cell also has a **receptive field** that is the region in stimulus space that evokes a response in that cell. The receptive field of a photoreceptor in the retina is a particular location in the visual space in front of the eye and a range of colors to which that receptor is sensitive. The receptive field for a somatosensory nerve in the skin is the area of skin that elicits a response. The receptive field for an olfactory neuron is the range of chemicals it can detect. Cells in the CNS concerned with sensory information also have receptive fields. Different cells handle sensory information from the feet than that from the hands. The incoming information arrives on "**labeled lines**"; the CNS processors know from whence it comes. There are several locations in the brain that have receptive fields including the same location in visual space. The receptive fields of these higher-order cells are more complex, as signal processing has occurred comparing the output from one lower order cell with that of others.

Mechanosensory **transduction** is direct, by mechanosensitive channels in the membrane. The sensory cell often has molecules or structures to focus the mechanical energy or filter out undesired mechanical disturbances, and there may be an elaborate organ—such as that comprising the outer, middle, and inner ear to deliver the desired mechanical energy to the appropriate cell. In the end, a relatively nonspecific cation channel opens and both Na and K ions move down their concentration gradients. In skin mechanoreceptors, such as the **Pacinian corpuscle** discussed below, there is a greater driving force on Na ions, so more Na than K moves and the cell depolarizes. The number of mechanosensitive channels that open is proportional to the amount the membrane is stretched by the stimulus. A larger stimulus will open more channels and produce a larger depolarization (Fig. 4-1). If the depolarization is large enough, action potentials will be initiated and will propagate toward the CNS.

The situation is more complex in the ear, because the sensory cells (called **sensory hair cells**, for the hair-like appearance of the modified cilia on their apical surface) are part of an epithelium that separates two different solutions. However, mechanical disturbance of these cells by the appropriate sound also leads to inward current through mechanosensitive channels and depolarizes the cell. The sensory hair cells are short and synapse with auditory nerve cells in the ear. The hair cells do not have action potentials; they are short compared to their length constant, so they can rely on passive spread to open Ca_V channels to release transmitters.

Some taste chemosensation is supported directly by chemosensitive channels, as in the glutamate receptors for the **umami** taste (the distinctive savory taste of glutamate); these are relatively nonselective cation channels that depolarize the cells. Others use channels even more directly; Na moving through epithelial sodium channels (ENaCs) depolarizes cells to provide the salty

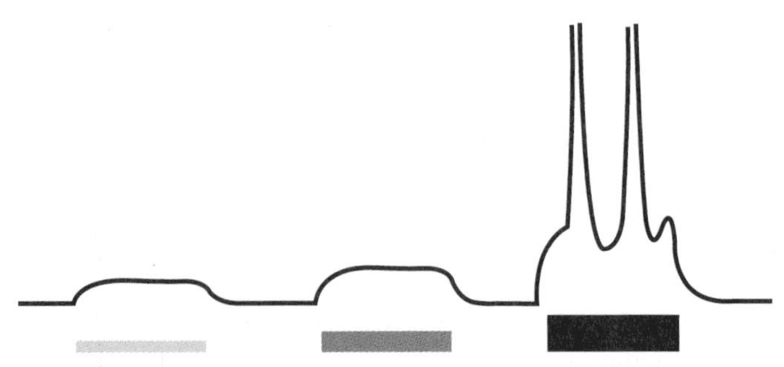

Figure 4-1. The changes in membrane potential of a mechanosensory nerve ending to stimuli of three different amplitudes.

taste sensation. Odors are detected by G protein–coupled receptors (GPCRs) whose G proteins activate adenylyl cyclase, thus elevating levels of cyclic adenosine monophosphate (cAMP). The cAMP opens a **cyclic nucleotide–gated** (CNG) nonspecific cation channel that depolarizes the cell. CNG channels are tetramers with six TM segments and are structurally similar to K_V channels but lack the latter's exquisite selectivity for K ions and the voltage sensitivity.

Light transduction also involves GPCRs with seven TM segments: rhodopsin in the rods and three other opsins in the cones tuned for short, medium, and long (or blue, green, and red) wavelengths. The chromophore that absorbs the light is 11-cis retinal (MW 284). Absorption of a photon triggers conversion of the retinal to the all-trans isomer, which causes a conformational change in the opsin protein informing the G protein that an event has taken place (Fig. 4-2). The G protein is called transducin; it was the first G protein to be identified and was named before the family was well known. Transducin activates a phosphodiesterase that hydrolyzes cyclic guanosine monophosphate (cGMP). In the dark, there is a CNG channel that is open and carrying inward current. The channel closes when the cGMP level drops; when the light is on, the dark current decreases and the cell hyperpolarizes. There is amplification along this

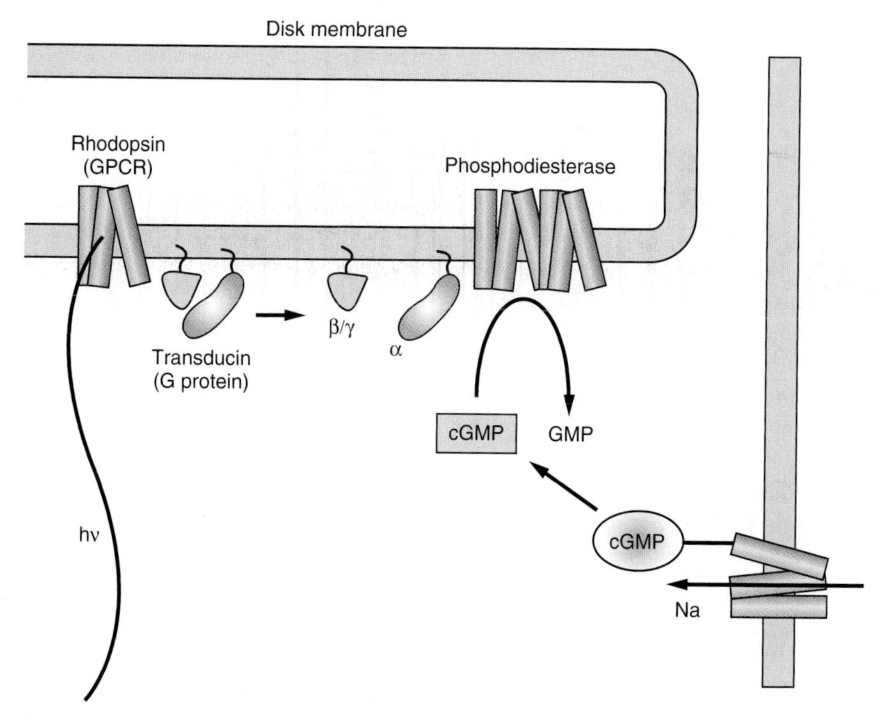

Figure 4-2. The processes linking light absorption by rhodopsin and the closing of cyclic nucleotide–gated channels.

chemical pathway, so one photon leads to the closure of many CNG channels. The hyperpolarization reduces a steady output of synaptic vesicles to pass the message on to the next cell in the pathway to the brain.

The sensation of uncomfortably hot skin temperature has been linked to the direct activation of a channel called VR1, for vanilloid receptor. It is also known as the capsaicin receptor because it can be activated by the vanilloid capsaicin, the major piquant ingredient in hot peppers. VR1 is a member of the transient receptor potentials (TRP) channel family; it has a six-TM architecture and is permeable to cations. Raising the temperature into the range of 42°C (107.6°F), which many human observers have identified as painfully hot, opens this channel, depolarizes the sensory ending, and initiates a train of action potentials. Other members of the TRP family have been associated with temperature sensation, though not all with pain.

(5) Pain is a condition with a long philosophical history. A few specific nociceptors have been identified, but there are also many other receptors that may be associated with pain. Elevated K^+ from damaged cells or the direct cutting of a nerve cell can induce action potentials that may be interpreted as pain. Acid-sensing ion channels (ASICs) in the ENaC family respond to lactic acid released in the heart and depolarize nerves that provide the sensory pathway for the painful experience of angina. P_2X_3 receptor channels, which can be activated by adenosine triphosphate (ATP) released by damaged cells, have been associated with pain from overstretched bladders, and P_2X_4 receptors have been associated with a neuropathic pain generated within the nervous system without obvious outside stimuli.

SENSORY ADAPTATION

(6) All senses except pain adapt; if they are presented with a maintained stimulus, the response will diminish in time. The Pacinian corpuscle adapts rapidly and responds to a sustained stimulus with only one or two action potentials at the start (Fig. 4-3). When the stimulus is released, there is an off-response and another action potential is initiated. Most of this adaptation takes place in the onion-like capsule of accessory cells that surrounds the nerve ending. When one side of the capsule is distorted by the stimulus, at first the distortion is transmitted to the nerve ending and the nerve depolarizes. Then the capsule balloons out to the sides, the forces on the nerve are relieved, and the nerve stops firing. When the stimulus is removed, the capsule rebounds to its original shape, transiently pushing the sides of the nerve in the process. The Pacinian corpuscle is tuned to provide maximum information about vibratory stimuli and to ignore steady pressure.

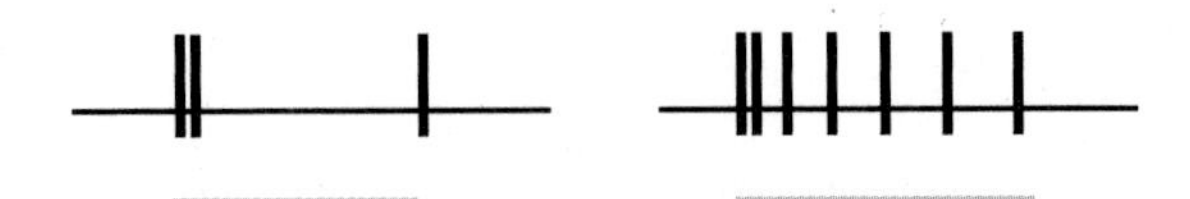

Figure 4-3. Fast and slow sensory adaptation.

Muscle spindle organs are sensory structures embedded in skeletal muscles, which provide information about the length of the muscle to the CNS (see Fig. 1-3). Muscle spindles adapt rapidly to changes in length but also continue to fire during a sustained stimulus. The firing rate decreases slowly during the stimulus; muscle spindles are said to be slowly adapting (Fig. 4-3).

The nervous system is usually more interested in changes in the environment, and by reducing the messages indicating that a stimulus is still present, more attention can be given to any changes. Adaptation takes place at many levels, from accessory tissue before the receptor potential, the receptor potential itself, the encoding mechanism that initiates action potentials, and at many higher synapses where the incoming message is integrated with other signals. Adaptation to light occurs by constricting the pupils, by photobleaching the pigments, and by feedback regulation of the steps in the biochemical cascade.

Many senses have some form of **efferent control**. The sympathetic nervous system can release norepinephrine onto the Pacinian corpuscle, which will increase its sensitivity to mechanical stimuli. Muscle spindle organs have efferent nerves (γ motor nerves) that set the range of lengths to which the sensory nerve is most sensitive. There are also motor hair cells in the ear that can selectively enhance the sensitivity of sensory hair cells to particular sounds. There are many controls on the eye to assure that the object of interest is suitably focused on an appropriate portion of the retina even as the head changes its position in space.

KEY CONCEPTS

Each sensory cell has an adequate stimulus.

Touch, hearing, and other mechanosensation occur via mechanosensitive channels.

Taste is mediated by chemosensitive channels and odor by G-protein–coupled receptors and cyclic nucleotide–gated (CNG) channels.

Vision is also mediated by G-protein–coupled receptors—e.g., rhodopsin—and CNG channels.

Pain is mediated by acid-sensing ion channels and purine-activated channels.

All senses except pain adapt.

STUDY QUESTIONS

4–1. *What is an adequate stimulus?*

4–2. *What is a receptive field?*

4–3. *What is sensory transduction?*

4–4. *Are receptor potentials always depolarizing (or receptor currents always inward)?*

4–5. *What is sensory adaptation and how/where does it arise?*

4–6. *What is a labeled line?*

SUGGESTED READINGS

Lledo PM, Gheusi G, Vincent JD. Information processing in the mammalian olfactory system. *Physiol Rev* 2005;85:281–317.

Macefield VG. Physiological characteristics of low-threshold mechanoreceptors in joints, muscle and skin in human subjects. *Clin Exp Pharmacol Physiol* 2005;32:135–144.

Matulef K, Zagotta WN. Cyclic nucleotide-gated ion channels. *Annu Rev Cell Dev Biol* 2003;19:23–44.

McCleskey EW, Gold MS. Ion channels of nociception. *Annu Rev Physio.* 1999;61:835–856.

Nicolson T. Fishing for key players in mechanotransduction. *Trends Neurosci* 2005;28:140–144.

Tominaga M, Caterina MJ. Thermosensation and pain. *J Neurobiol* 2004;61:3–12.

Action Potentials

5

OBJECTIVES

- *Describe the activation of action potentials.*
- *Explain the propagation of action potentials.*
- *Describe the membrane currents underlying action potentials.*
- *Describe the activity of channels producing action potentials.*
- *Explain the membrane basis of the action potential threshold and refractory period.*
- *Explain actions of calcium, local anesthetics, and neurotoxins on action potentials.*
- *Describe the relationship between channel activity and cardiac muscle contraction.*
- *Describe the membrane basis of intrinsic cardiac pacemakers.*
- *Describe the effects of acetylcholine and NE on cardiac action potentials.*

ROLE OF VOLTAGE-SENSITIVE SODIUM CHANNELS

Action potentials are changes in membrane potential that propagate along the surface of excitable cells. They are best known in nerve and muscle cells but also occur in some other cells, including egg cells associated with fertilization and plant cells. Unlike some other changes in membrane potential, action potentials are characterized as being "all-or-none"; they have a threshold for excitation and a stereotyped duration. Immediately following an action potential, the excitable cell has a refractory period when it is more difficult or impossible to elicit a second action potential.

Like most changes in membrane potential, action potentials are the result of changes in membrane permeability due to the activity of channels, or proteins embedded in the membrane lipid bilayer that facilitate the passive movement of specific ions down their electrochemical gradients. An action potential is a change in membrane potential from a resting potential of about −70 mV (the inside of the cell is negative) to about +30 mV and then back to the resting potential. Their duration in nerve and skeletal muscles is on the order of 1 ms; in ventricular muscle cells, their duration is several hundred milliseconds. In nerve and skeletal muscles, the underlying permeability changes are a transient increase in sodium permeability

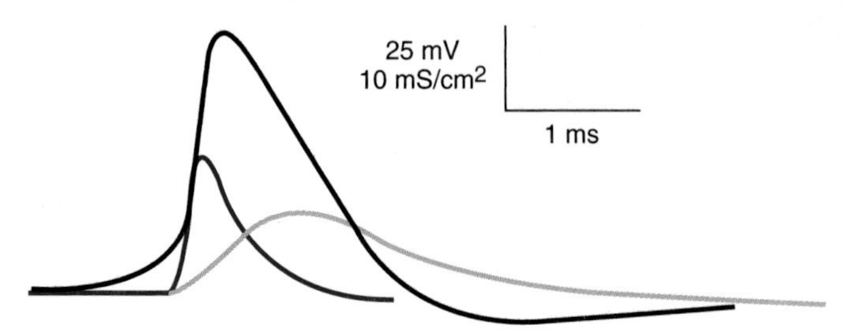

Figure 5-1. An action potential and the underlying changes in membrane conductance for Na and K.

followed, after a delay, by an increase in potassium permeability due, respectively, to the activation of sodium and potassium channels (Fig. 5-1). Cardiac action potentials are more complex and also involve the activation of calcium channels.

Action potentials are all-or-none and propagate because the sodium channels are voltage-sensitive. Depolarization, the reduction of the membrane potential, from –70 mV toward 0 mV, induces a conformational change within a few hundred microseconds in the sodium channel protein, which leads to in increase in permeability to sodium ions. Sodium ions move into the cell through these open channels and bring positive charge with them, which further depolarizes the cell, opening more sodium channels (Fig. 5-2).

This positive feedback loop persists until all of the sodium channels have opened. In life, once the loop is started, it continues to completion. The depolarization spreads passively to adjacent regions of the membrane and activates nearby sodium channels. This wave of molecular conformational change and electrical activity propagates over the length or surface of the cell at velocities up to 120 m/s.

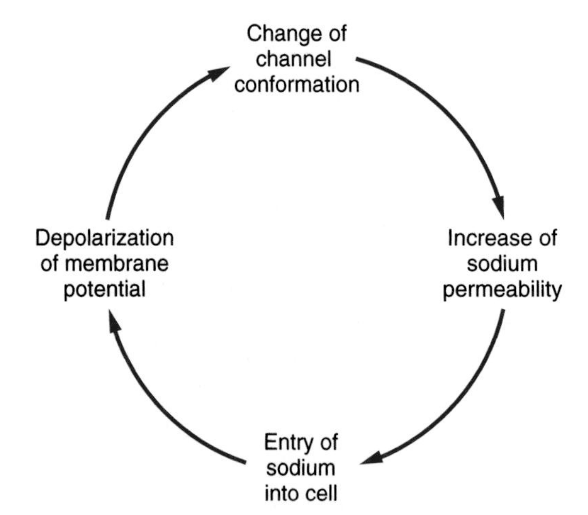

Figure 5-2. The action potential's positive feedback cycle.

Potential energy that is stored in the sodium concentration gradient is sequentially used along the propagation path. The conduction velocity is determined by the rate of molecular change and the electrical properties of the cell that control the spread of potential changes (cable properties).

About one millisecond later, the sodium channels undergo a second conformational change and inactivate. In this third conformation, they are closed and sodium no longer passes through. In addition, the Na channels are unable to open again until the membrane is repolarized back to the resting potential for a few milliseconds to allow recovery from inactivation (Fig. 5-3). This automatic closing of the sodium channels limits the duration of nerve and skeletal muscle action potentials. Loss of the ability to open again produces the refractory period.

The outward movement of K ions carrying positive charge out of the cell produces the repolarization (the falling phase of the action potential). In some cells, K_V channels—whose activation is slower than that of sodium channels—facilitate repolarization. In mammalian myelinated axons, the repolarizing current passes through the (non-voltage-sensitive) potassium channels that

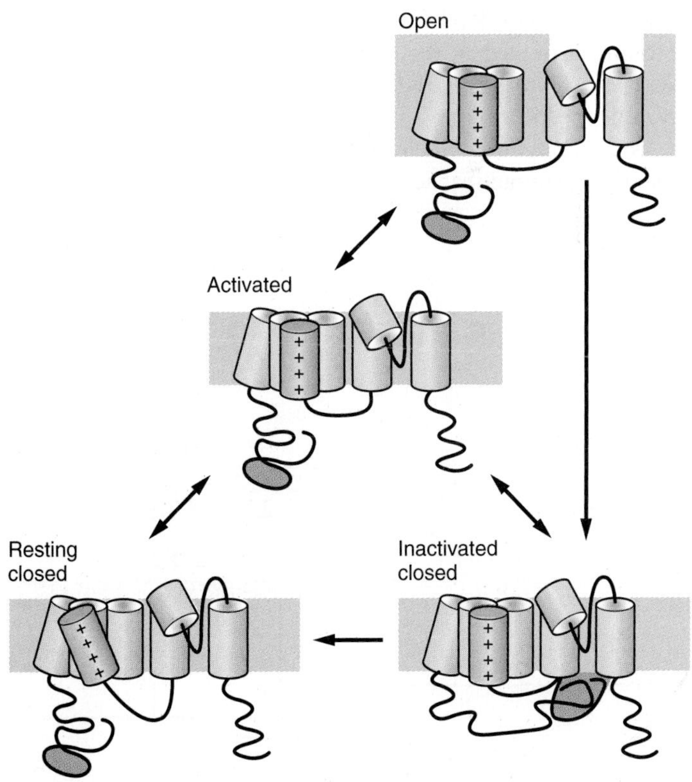

Figure 5-3. Sodium channels can be in different functional states.

produce the resting potential. The axons seem to be an exception; the presynaptic nerve terminals and the cell bodies of most neurons have K_V channels.

VOLTAGE CLAMPING

This understanding of the action potential mechanism comes from the work of Alan Hodgkin and Andrew Huxley about 50 years ago. Working with giant nerve axons isolated from squid, they were able to build electrodes and electronic circuitry that let them break the positive feedback loop and measure the effect of a change in membrane potential in the ionic permeabilities without any change to the membrane potential due to the movement of ions. Their technique was to include the nerve membrane in a negative feedback circuit (Fig. 5-4).

A pair of electrodes measures the membrane potential; this is then compared with a desired command potential. If the membrane potential is different than the command potential, a current is made to flow through the membrane in a direction that reduces the difference. This is analogous to a thermostat that measures the temperature and turns on heating or cooling if the measured temperature is different from the desired value.

Hodgkin and Huxley used square pulses for their command potential; the membrane potential was changed in a few microseconds from the resting potential and then held at a constant level for several milliseconds while the membrane currents were recorded. After that, the membrane potential was returned to the resting level. When the pulse is from the resting potential to 0 mV, four different kinds of current can be identified (Fig. 5-5).

The first is the charge movement necessary to change the potential or change the charge on the membrane capacitance. Second, there is a small outward current

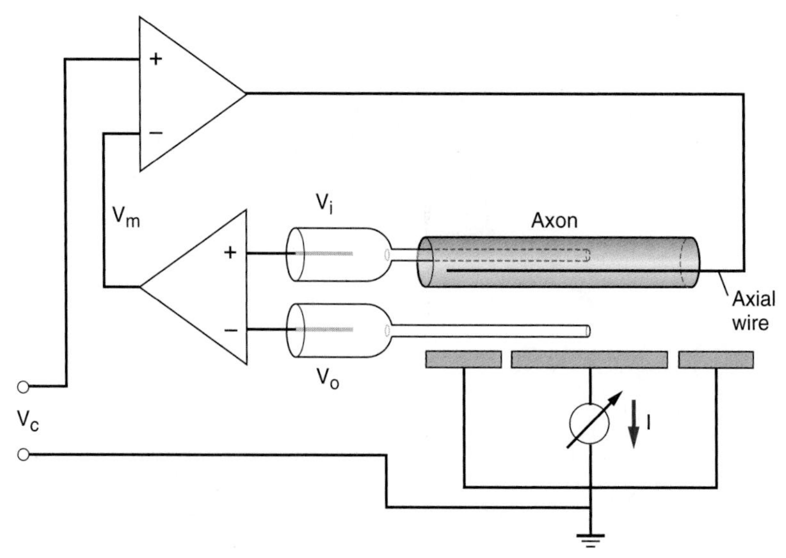

Figure 5-4. A simplified voltage-clamp circuit for a squid giant axon.

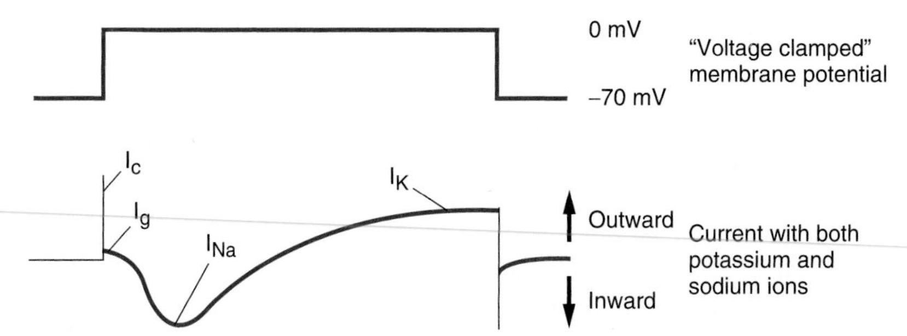

Figure 5-5. The membrane currents in response to a voltage clamp pulse. I_c, capacity current; I_g, gating current; I_{Na}, sodium current; I_K, potassium current.

called the "gating" current. Then there is an inward current that is replaced in a few milliseconds by an outward current, which lasts as long as the pulse.

One can replace the contents of a segment of squid axon with a simple salt solution and maintain functioning channels. By changing the solutions bathing both sides of the membrane, it has been possible to separate the currents carried by Na ions (I_{Na}) and K ions (I_K) and also to see the gating currents (I_g) still present in the absence of either ion (Fig. 5-6). Choline or tetramethylammonium ions can be used to maintain conductivity in the solutions while not permeating the membrane channels.

Notice that at this potential, the Na current is inward and the K current outward. The Na current activates or increases more rapidly than the K current.

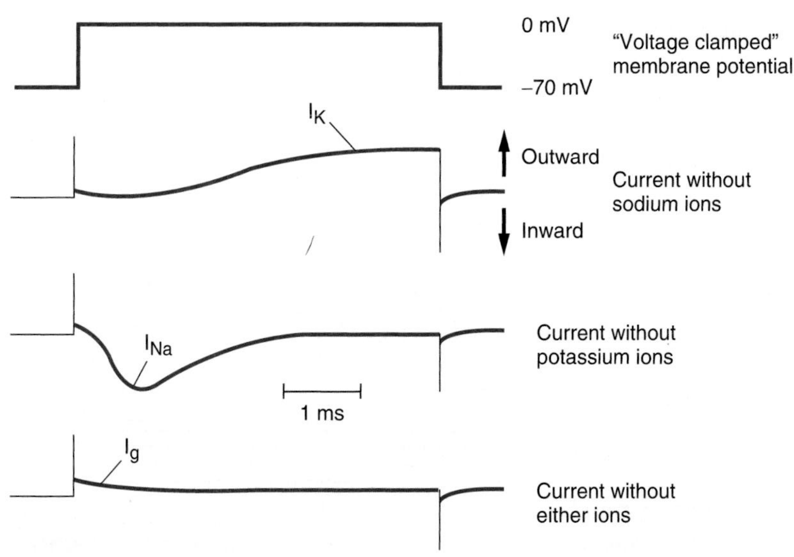

Figure 5-6. The separation of the currents by changing the solutions.

The Na current inactivates or decreases during the pulse, even though the membrane potential is kept at 0 mV, whereas the K current remains for the duration of the pulse.

If the membrane potential is pulsed from −70 to −140 mV (a hyperpolarizing pulse), the gating current and ionic currents do not appear; only the capacity transient is seen. If the potential is pulsed to other depolarized potentials, all four components of the current are present, although their amplitude and time course and, in the case of I_{Na}, direction may change (Fig. 5-7).

The Na current becomes more inward between the resting potential and about 0 mV. Larger pulses produce less inward Na current until, at about +60 mV, no net current passes through the Na channels. Still larger pulses can drive outward Na current through the Na channels. The reversal of the current occurs at the sodium equilibrium potential, E_{Na}. If the ratio of the sodium concentrations bathing both sides of the membrane are changed, this reversal potential also changes. With modest depolarizations, the inward current increases because larger pulses open more sodium channels. However, the less negative potential decreases the inward driving force on the sodium ions; after most of the Na channels have been opened, still larger depolarizations decrease the Na current. When the membrane potential exceeds the sodium equilibrium potential, Na is forced out of the cell through the open Na channels. In a free-running action potential, the membrane potential never exceeds the sodium equilibrium potential and there is always a net entry of Na into the cell.

The Na current activates and inactivates more rapidly as the size of the pulse is increased. If a second pulse is given immediately after the first, the gating current and the sodium current during the second pulse are smaller than during the first pulse (Fig. 5-8). They both recover in parallel as the duration between pulses is increased. The rate of recovery from inactivation is also voltage-dependent, as the channels recover more rapidly at more hyperpolarized potentials.

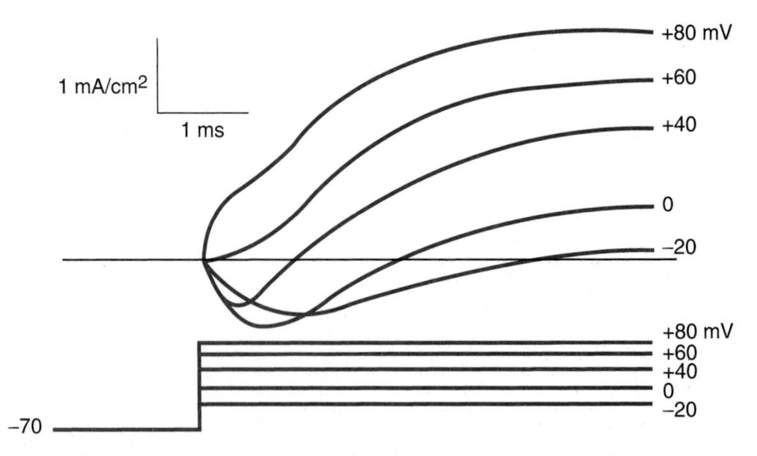

Figure 5-7. The current's responses to voltage steps of varying amplitude. Capacity current transients not shown.

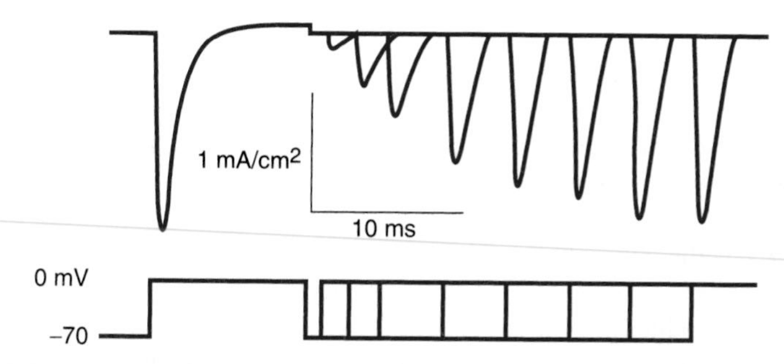

Figure 5-8. The recovery from inactivation shown by a two-pulse experiment with different amounts of time at the resting potential between pulses. Capacity current transients not shown.

The K current increases and becomes more rapid as the membrane potential is increased. Above about +20 mV, the increase in amplitude becomes proportional to the change in potential, indicating that all of the channels are open and that only the driving force continues to increase.

The gating current is a direct sign of the conformational changes in the sodium channel proteins. These molecules contain charged groups and dipoles that move or reorient when the electrical field changes. This movement can be measured as the gating current. As the pulse is made progressively more positive and more sodium channels open, the amplitude of the gating current grows and the currents become more rapid. Above about +20 mV, these two changes are complementary and the area under the gating current trace is constant, indicating that all of the channels are undergoing conformational changes and are doing so more rapidly at more positive potentials.

The capacitance current increases linearly with the size of the pulse because it requires more charge to change the voltage by larger amounts.

Hodgkin and Huxley separated the currents and showed how the ionic currents were proportional to the driving force on the ions. Then they created mathematical equations that emulated the amplitude and time course of the permeability changes and showed that these equations could predict the amplitude and time course of action potentials as well as their threshold, conduction velocity, refractory period, and several other features. Their concept of describing ionic current as the product of conductance times driving force has been used to describe most of the remaining electrophysiological phenomena in all cells and tissues.

The Hodgkin-Huxley equations are available in a commercial computer program called Neuron. The website (http://pb010.anes.ucla.edu/VC.htm) has a JavaScript rendition that will allow you to manipulate the equations with most modern web browsers.

THRESHOLD

The threshold arises because there are two different effects of small depolarizations. On the one hand, depolarization will increase the probability that voltage-dependent sodium channels open and permit inward current, which will lead to further depolarization (Fig. 5-2). On the other hand, depolarization moves the membrane potential further away from the potassium equilibrium potential, increasing the net driving force on potassium ions and thus producing an outward current through the resting potential potassium channels, which will lead to repolarization.

If a sufficient number of sodium channels are opened so that the inward sodium current exceeds the outward potassium current, the cell has exceeded threshold and will continue to depolarize until all of the available sodium channels have opened. Treatments that reduce the sodium current—e.g., reducing extracellular sodium concentration or reducing the number of sodium channels—will elevate the threshold.

REFRACTORY PERIODS

During an action potential, most of the Na channels activate or open and then inactivate and close into a state that differs from their condition before the action potential. In order to recover from inactivation and be available to open again, the Na channels must spend some time with the membrane potential near the resting potential. They will not recover if the membrane remains depolarized.

8 During this recovery, the axon is said to be refractory because it is resistant to stimulation. The refractory period is divided into two segments: an absolute refractory period when no stimulus, however large, can elicit a second action potential, followed by a relative refractory period when the axon can be stimulated again but requires a larger stimulus to elicit the second response that was needed for the first (Fig. 5-9).

During the absolute refractory period so few sodium channels have recovered that even if all of the recovered channels were opened, there would be insufficient sodium current to exceed the outward potassium current, which tends to restore and maintain the resting potential. During the relative refractory period, a larger depolarization is required because a larger fraction of the available sodium channels

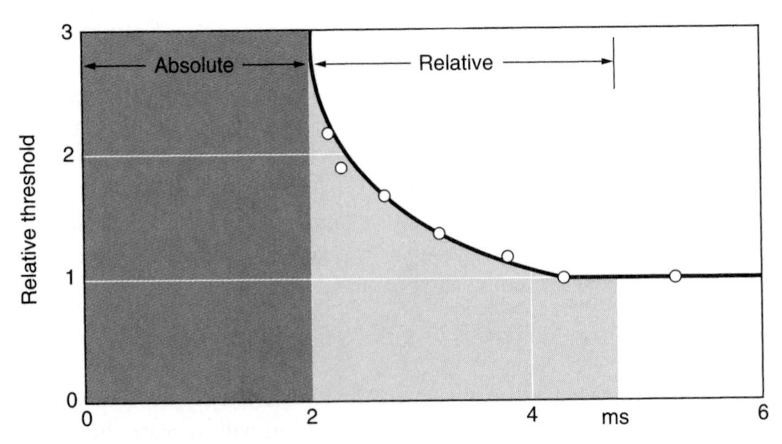

Figure 5-9. The absolute and relative refractory periods.

must be opened to obtain the same number of channels opened in the first stimulus. In addition, in many nerve and muscle cells, there are more open potassium channels immediately following an action potential, which also makes the cell more difficult to excite a second time.

MYELINATION

Vertebrate nervous systems present a specialization of nervous function not seen in invertebrates, namely myelination (Fig. 5-10). Accessory cells wrap nerve axons with many layers of their own membrane, electrically insulating most of the cell. Sodium channels cluster in the regions between these wraps, in the **nodes of Ranvier**. The Na current enters the cell only at these nodes; excitation "jumps" from node to node in what is called **saltatory conduction**. The spread between nodes is the same passive spread seen in unmyelinated nerve cells, but it is more effective; that is, it produces a more rapid conduction velocity. The myelin wraps increase the resistance between the axoplasm and the surrounding media, which, in turn, increases the length constant for passive spread. The myelin also increases the effective thickness, which decreases the effective capacitance and reduces the amount of charge required to change the potential. Both effects speed conduction.

DISEASES

There are many diseases or conditions of reduced or excessive excitation of cells. Perhaps the most familiar is the conduction of acute pain information, which is frequently treated with local anesthetics; these act by blocking the Na_V channels. Some forms of epilepsy and some cardiac arrhythmias are also treated with Na channel blockers. One type of **long-QT** (LQT) **syndrome** has been linked to a mutation in one of the Na channel genes, and a **hypokalemic periodic paralysis** has been linked to another. Other LQT syndromes have been associated with K_V channels.

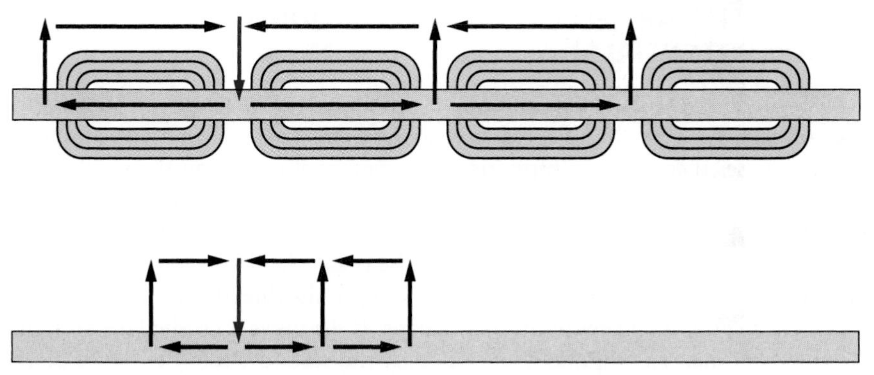

Figure 5-10. The effect of myelination on the longitudinal spread of current.

Hypocalcemia is associated with increased excitability of nerves and skeletal muscle and may produce uncontrollable muscle contraction (**tetany**). Hypercalcemia renders nerves and muscles less excitable. Calcium is thought to bind to the membrane near the voltage sensor of the sodium channel and have an effect on the channel protein similar to hyperpolarization. The voltage sensitivity of the sodium channels is "shifted" along the voltage axis by changes in extracellular calcium. The result is that, in low calcium conditions, the sodium channel opens in response to a smaller stimulus or even spontaneously at the resting potential. The calcium binding does not change the resting potential as measured with electrodes in the bulk compartments on both sides of the membrane.

There are **demyelinating** diseases, such as **multiple sclerosis** (MS), where myelin is lost and conduction can become slower or fail altogether. MS is an autoimmune disease and is generally treated with steroids such as prednisone. The symptoms can be eased by providing air-conditioning or moving to a cooler climate. Cooling helps, somewhat paradoxically, because although it slows the opening of sodium channels and thereby slows the propagation velocity, cooling also slows the inactivation of Na_V channels and increases the duration of the action potentials; thus the greater Na influx makes the propagation more reliable.

DRUGS AND TOXINS

After the identification of these specific Na and K conductances, they were shown to be molecularly separate because they differ in pharmacology and respond differently to various drugs. **Tetrodotoxin** (TTX), a poison found in the internal organs of puffer fish, selectively blocks nerve Na_V channels at nanomolar concentrations. Local anesthetics such as **lidocaine** or benzocaine also block Na_V channels. There is a greater diversity among K_V channels and also among the drugs that block them. Tetraethyl ammonium (TEA) ions and 4-aminopyridine are among the K_V channel blockers. There are also compounds that chronically activate Na_V channels, such as veratridine, pyrethroid insecticides, and brevetoxin, one of the red-tide toxins.

EXTRACELLULAR RECORDINGS—COMPOUND ACTION POTENTIALS

Action potentials can be recorded with a pair of wires placed on the surface of a nerve bundle, typically about 1 cm apart. When a nerve impulse passes these wires, a **biphasic** action potential is seen on the oscilloscope screen (Fig. 5-11). This is a differential recording of the same nerve impulse that would appear as in Fig. 5-1 if the recording were made with an intracellular microelectrode. One deflection occurs as the impulse passes the first wire and the second occurs as it passes the second wire. They are in opposite directions because the two wires lead to opposite plates of the oscilloscope. If the nerve is crushed between the electrodes so that the impulse does not reach the second electrode, the response becomes **monophasic**.

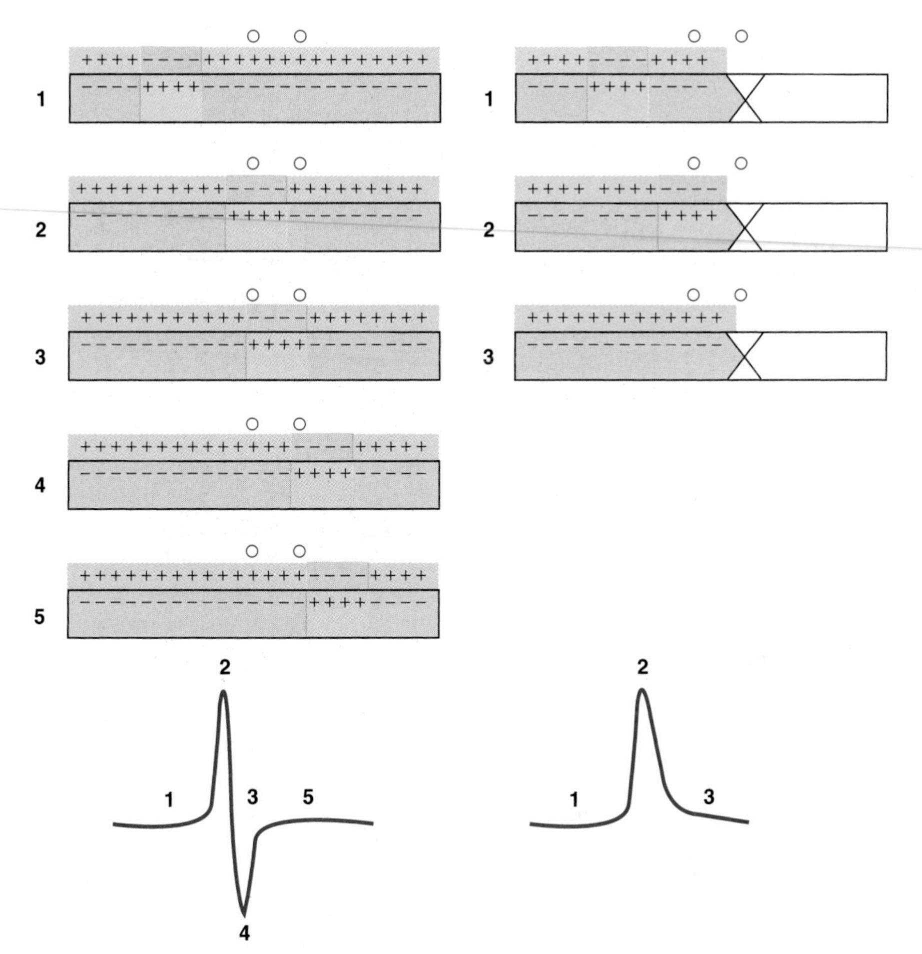

Figure 5-11. Externally recorded action potentials. Left, biphasic action potential recorded from an intact axon. Right, monophasic action potential recorded near the site of a crush injury.

This type of recording with external electrodes is used clinically to test nerve integrity. A nerve bundle can also be stimulated with another pair of wires on a remote stretch of the same bundle. With appropriate equipment, stimulation and recording can be made through the skin without dissecting out the nerve bundle. When a nerve bundle is stimulated, more than one axon may be excited. The electrical recording of the combination of the action potentials produced is called a **compound action potential**. The compound action potential is also biphasic if the nerve is intact between the recording wires.

Besides being biphasic, there are many differences among compound action potentials recorded with external electrodes, the single-cell action potential recorded with an electrode inside the cell, and a reference electrode outside the cell.

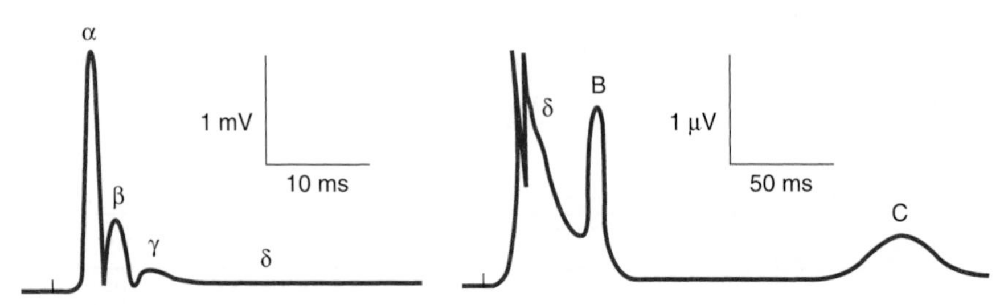

Figure 5-12. A compound action potential. Left, high sweep speed. Right, lower sweep speed, higher vertical gain.

The compound action potentials are much smaller, on the order of 1 mV, and there is no sign of the resting potential because both wires are outside the nerve. The compound action potential is not all-or-none because a larger stimulus will bring more individual axons above threshold and the compound action potential's amplitude is proportional to the number of axons firing. The compound action potential becomes smaller and longer at greater distances from the stimulating electrodes because the conduction velocity of the various axons is not exactly the same and the action potentials disperse as they travel away from the stimulation site.

The threshold and conduction velocity of the various axons within a nerve bundle vary with the diameter of the axons. Large axons have a lower threshold to stimulation by external electrodes. (Of course, in life they are usually stimulated more selectively by a specific receptor or synaptic input.) The larger-diameter fibers have a lower threshold; more of the stimulating current flows through them because they have a lower internal resistance. Larger axons also have a more rapid conduction velocity, again because of their lower internal resistance.

Vertebrate peripheral axons are classified by their diameter (or conduction velocity or threshold to external stimulation). There are groups of nerve fibers with similar diameters. The groups of different diameter can be distinguished as separate elevations in the compound action potential (Fig. 5-12). There is some correlation of function with diameter. For example, large myelinated motoneurons leading to skeletal muscles are Aα fibers and small unmyelinated fibers carrying pain information are C fibers. The larger fibers have faster conduction velocities and lower thresholds to external electrical stimuli.

CARDIAC ACTION POTENTIALS

The heart is a pump made up of excitable muscle cells. The electrical activity of these cells controls their contraction. The overall control of the heart's pattern of contraction is by the spread of action potentials through a special conducting system of modified heart muscle cells (**Purkinje fibers**) and through the atrial and ventricular muscle cells themselves (Fig. 5-13).

Normally, **pacemaker** cells in the **sinoatrial** (SA) **node** control the heart rate. These **SA nodal** cells automatically produce nodal action potentials 60 to 80 times per minute. Current flows from the pacemaker cells into

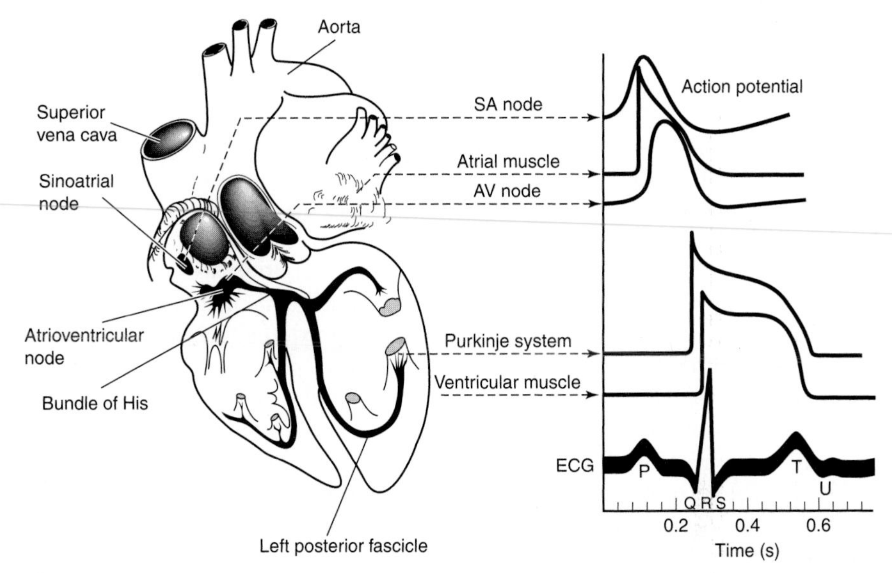

Figure 5-13. Action potentials from different regions of the heart. (Modified from Ganong WF. *Review of Medical Physiology*, 22d ed. New York: Lange Medical Books/McGraw-Hill, 2005, with permission.)

atrial muscle cells via gap junctions. In the atrial cells, this current initiates muscle action potentials that propagate over the atria from cell to cell via gap junctions. At the **atrioventricular** (AV) border, excitation spreads through gap junctions and excites nodal action potentials in **AV nodal** cells. Action potentials propagate through the AV node and then initiate muscle-type action potentials in the Purkinje fibers. These action potentials travel to various locations in the ventricles and then spread to ventricular muscle cells, which also have fast action potentials.

The muscle and nodal action potentials are distinguished by their rate of depolarization and their conduction velocity. The muscle action potentials are found in the atrial and ventricular muscle cells and the Purkinje fibers. The nodal action potentials are normally found in the SA node and the AV node. The muscle action potentials have long durations, so the first atrial cell that started its action potential is still depolarized after the last atrial muscle cell is excited. Similarly, there is a period when all of the ventricular muscle cells are depolarized.

The **electrocardiogram** (ECG) is an indication of these action potentials seen from electrodes on the surface of the body. The **P** wave corresponds to the spread of the start of the action potentials over the atria. The **QRS** wave indicates the excitation of the ventricular muscle cells, and the T wave represents the repolarization of the ventricular muscle cells. The **QT interval** is the average duration of the ventricular muscle action potentials. Repolarization of the atria occurs during the QRS interval and is not normally visible. The action potentials in the nodal tissue and the Purkinje fibers are not visible in the standard ECG. Intracardiac ECG recordings have been made of bundle of His action potentials.

CARDIAC MUSCLE ACTION POTENTIALS

In cardiac muscle action potentials, current from adjacent cells depolarize the cell to a level where fast, voltage-dependent Na channels (Na_V) open and rapidly depolarize the membrane toward the sodium equilibrium potential (phase 0 in Fig. 5-14). These channels are similar to the sodium channels of nerve and skeletal muscle; they open in response to depolarization. They are also blocked by local anesthetics. After opening, they inactivate quickly and the membrane potential starts to return. However, the depolarization also opens voltage-activated L-type Ca_V channels that do not inactivate. This maintains the action potential in the **plateau** phase (phase 2). Reducing external Ca or adding drugs that block calcium channels will reduce the plateau phase and also reduce the strength of muscle contraction. Cardiac muscle, unlike skeletal muscle, requires external Ca for contraction.

Cardiac muscle cells also differ from nerve and skeletal muscle by lacking the fast K channel for quick repolarization. The potassium conductance system of the heart is rather complex; at least five different components have been identified on the basis of their kinetics and voltage dependence. Two of these are important to understand the plateau phase. During the plateau phase, the conductance is less than during **diastole**, the period between action potentials. This is because of the inward rectifier channel (K_{ir}), which is responsible for maintaining the resting potential and has a high conductance near and below the resting potential (at more negative potentials); it does not conduct during the plateau phase when the membrane is depolarized.

The K_{ir} channel rectifies, allowing current to flow and maintain the resting potential, but it does not allow much current to flow out during depolarization. The rectification is caused by Mg^{2+} or other polyvalent cations from the internal

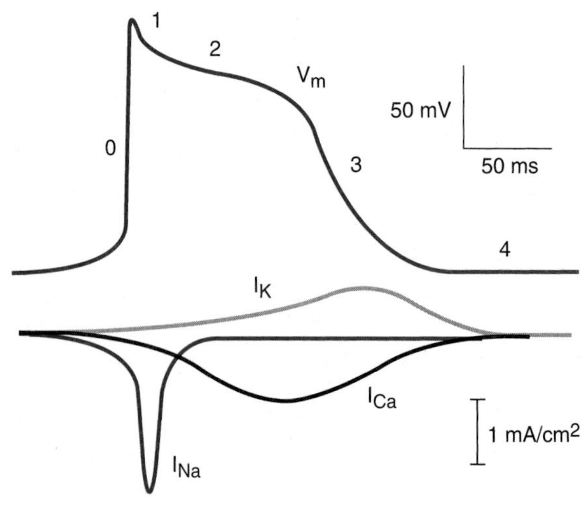

Figure 5-14. A ventricular muscle cell action potential and its underlying ionic currents.

solution moving into the channel and plugging it when the cell is depolarized. The low conductance to K during the plateau phase means that the modest conductance to Ca^{2+} through the Ca_V channels maintains the membrane potential at depolarized levels during the plateau.

Slow K_V channels open very slowly during the action potential and are responsible for the downward slope during the plateau phase. When the membrane potential falls below a certain level, the Ca_V channels close and the repolarization toward the potassium equilibrium potential accelerates (phase 3). Since the membrane is no longer depolarized, the K channels close.

The description above is a simplified view of cardiac muscle action potentials. The complete story has several more K channels and must account for differences among muscle action potentials in different regions of the heart as well as age-related changes. There are two K_V channels that open transiently just after the Na_V channels and produce the initial partial repolarization (phase 1) from the peak to the plateau (IK_{to}). There are at least two different slow voltage-dependent K channels with similar kinetics but distinct pharmacology (IK_R and IK_S). Some cardiac muscle cells have T-type calcium channels. In all cardiac cells, some current is carried by the sodium-calcium exchanger and by the Na/K pump.

The regional and age-related differences in the action potentials are functionally and clinically important. The ventricular muscle's action potentials near the **endocardial** (inner) surface have a longer duration than those near the **epicardial** (outer) surface. More work is done by the inner fibers, and they are more likely to be damaged in a heart attack. These differences must arise because of a different balance of Na, Ca, and K channel activities. The interactions between the effects of different channels are complex and are best explored with computer models. Ventricular muscle action potentials also vary between different animal species, which further complicates the research. Clearly more research is necessary to understand the details.

SA AND AV NODE ACTION POTENTIALS

15

The overall control of the heart's pattern of contraction is normally initiated by action potentials that spontaneously arise 60 to 80 times per minute from modified muscle cells in the SA node. Similar action potentials are also seen in the atrioventricular (AV) node, where they regulate the activation of the ventricles. The AV node's cells are capable of spontaneous activity of about 40 action potentials per minute; in normal hearts, however, the atrial cells drive them at the rate set by the SA node.

The action potentials in the nodes lack the rapid upstroke and do not have as pronounced a plateau phase as the cardiac muscle action potentials. They are further characterized by the slow depolarization between action potentials: the **pacemaker potential**. These cells fire rhythmically; they are never at rest and have no true resting potential.

The upstroke of the action potential is produced by a slow inward current carried primarily by Ca (Fig. 5-15). There is an initial phase through T-type Ca_V channels and a major phase through L-type Ca_V channels. The T-type channels are

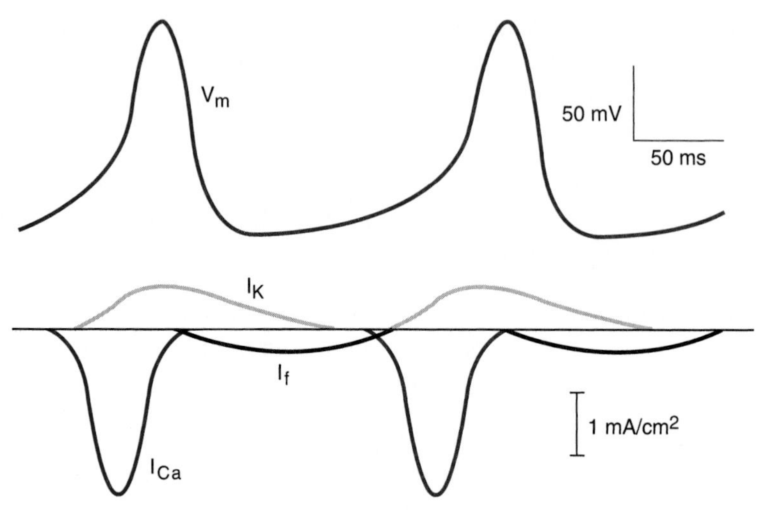

Figure 5-15. The SA node's action potentials and their underlying currents.

transient and have a low threshold for opening, near −60 mV. The L-type channels are long-lasting and have a higher threshold, near −30 mV. The L-type channels are similar to the Ca_V channels that maintain the plateau of the cardiac muscle action potentials; they are blocked by **dihydropyridines**. T-type channels have a different pharmacology. Reducing external Ca or adding Ca channel blockers reduces the amplitude of the node's action potentials. Outward K current gradually replaces the slow inward current and the cells repolarize towards E_K. As the potential passes −50 mV, an inward **hyperpolarization-activated** current, I_f, appears, competes with I_K, and eventually begins to depolarize the cell again. I_f is carried mainly by sodium ions. When the potential again passes −60 mV, the Ca_V channels are again activated and the cycle is repeated.

EFFECTS OF SYMPATHETIC AND PARASYMPATHETIC INNERVATION

The heart can beat spontaneously without nervous input. In normal life, however, the heart rate and its strength of contraction are regulated by the autonomic nervous system and circulating hormone levels. The autonomic nervous system controls many internal organs through its two divisions, the sympathetic and the parasympathetic nervous systems. These release **norepinephrine** (NE) and **acetylcholine** (ACh), respectively, onto the heart. The autonomic nervous system can also cause the adrenal medulla to release **epinephrine** into the blood. Epinephrine has effects on the heart similar to those of NE. Some of the details about the autonomic synapses and their pharmacology are described in the next chapter.

The cells in the SA and AV node cells have GPCRs that produce a stimulation (via $G\alpha_s$) or inhibition (via $G\alpha_i$) of adenylyl cyclase, which, in turn, raises or lowers cAMP levels in response to NE and ACh, respectively.

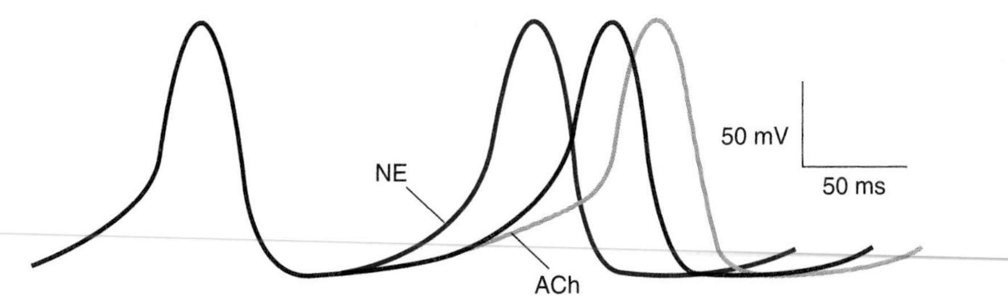

Figure 5-16. The effects of acetylcholine and norepinephrine on the SA node's action potentials.

The cAMP enhances the activity of the I_f channels. The end result is that NE increases the I_f current and thus depolarizes the cells more rapidly and increases the heart rate. ACh reduces the I_f current, slows the rate of depolarization, and reduces the heart rate (Fig. 5-16). The SA node controls the firing rate of the AV node. The effects on I_f lead to a speeding or slowing of conduction through the AV node. This can be detected clinically by observing in the ECG the time between atrial depolarization and ventricular depolarization, or the **PR interval**.

High ACh levels lead to the opening of another potassium channel (K_{ACh}). (It is a G-protein–activated inward rectifier GIRK channel.) This channel further reduces the tendency to depolarize between action potentials and can temporarily stop the heart.

Norepinephrine Also Increases Contractility

17 In the presence of NE, the plateau of the muscle action potentials is elevated and has a shorter duration (Fig. 5-17). This shortening of the action potential shortens the duration of the muscle contraction, which is functionally important for the heart. At high heart rates, the time required to refill the heart limits its performance. By reducing the time that muscle force is being generated (systole), more time is left for filling (diastole). The shortening of the ventricular action potentials can be seen in the ECG as a shortening of the QT interval.

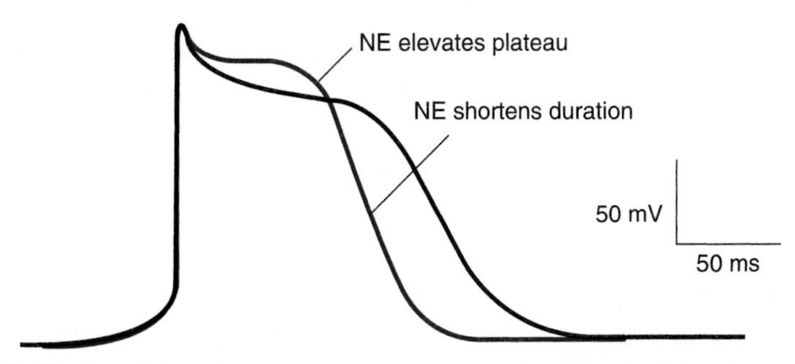

Figure 5-17. The effects of norepinephrine on ventricular muscle cells' action potentials.

NE increases the amplitude of the plateau by causing the action potential to open more L-type Ca channels. This drives the membrane closer to the Ca equilibrium potential. The increased Ca influx leads to a greater strength of contraction by a mechanism described in Chap. 7. NE shortens the duration by making the K_V channels open more rapidly. The effects on the K and Ca channels are mediated via cAMP acting as a second messenger, stimulating phosphokinase A (PKA) and phosphorylating the channels. This pathway also enhances the calcium reuptake mechanism by phosphorylating phospholamban. This speeds up muscle relaxation, as described more fully in Chap. 7.

Acetylcholine Reduces Atrial Contractility

The ACh-activated K channel (K_{ACh}) remains open during the action potentials; in atrial muscle and Purkinje fibers, it makes the plateau phase shorter and lower. The atrial contractions are weaker. ACh receptors are relatively sparse on ventricular muscle cells.

KEY CONCEPTS

Depolarization opens Na_V channels, which allows Na to rush in and produce further depolarization. This positive feedback loop produces the all-or-none quality and the propagation of action potentials.

K leaving the cell repolarizes the membrane potential and terminates action potentials.

Voltage clamping, or negative feedback control of the membrane potential, facilitates understanding of the currents underlying the action potential.

The amplitude and direction of the sodium current vary with the amplitude of voltage-clamp steps in membrane potential.

Depolarizing steps first activate and then inactivate Na current. They also activate K current after a delay.

The gating current is a direct sign of the conformational changes in the sodium channel proteins.

There is a threshold for action potential initiation.

Following an action potential, excitable cells have an absolute refractory period when they will not produce a second action potential and then a relative refractory period when a larger stimulus is required to produce a second action potential.

Myelination increases conduction velocity by increasing the length constant.

Hypocalcemia (low extracellular calcium) makes excitable cells more excitable.

Demylinating diseases slow the conduction velocity and may block the propagation of action potentials.

Action potentials appear differently when they are recorded with a pair of wires placed on the outside of a nerve bundle. Compound action potentials, the sum of many externally recorded action potentials, have properties that differ from those of single action potentials recorded with intracellular electrodes.

In the heart, action potentials arise automatically in the SA node and then spread from cell to cell over the heart via gap junctions.

Cardiac muscle cells have K_{IR} channels to maintain the resting potential, Na_V channels for the upstroke of the action potential, Ca_V channels for the plateau phase, and slow K_V channels for the repolarization.

SA node cells use Ca_V channels for the upstroke of the action potential, K_V channels for the repolarization, and a hyperpolarization-activated I_f channel to produce the slow "pacemaker" depolarization between action potentials.

Acetylcholine and NE slow or speed the heart rate, respectively, via G-protein–coupled receptors, which lead to a decrease or increase in I_f.

NE increases the amplitude of the plateau and decreases the duration of ventricular muscle action potentials.

STUDY QUESTIONS

5–1. At threshold, what ionic currents are balanced?

5–2. What are the effects of the following on conduction velocity?

a. Increasing axon diameter
b. Increasing internal resistance
c. Decreasing external sodium
d. Cooling
e. Demyelination

5–3. Experimentally, what are the absolute and relative refractory periods? How are they explained in terms of channels?

5–4. What are cable properties?

5–5. Draw an action potential and the underlying permeability changes for sodium and potassium.

SUGGESTED READINGS

Bezanilla F. Voltage sensor movements. *J Gen Physiol* 2002;120:465–473.

Goldin AL. Mechanisms of sodium channel inactivation. *Curr Opin Neurobiol* 2003;13:284–290.

Head C, Gardiner M. Paroxysms of excitement: Sodium channel dysfunction in heart and brain. *Bioessays* 2003;25:981–993.

Hille B. *Ion Channels of Excitable Membranes.* Sunderland, MA: Sinauer, 2001.

Hodgkin, AL. *The Conduction of the Nervous Impulse.* Springfield IL: Charles C Thomas, 1964.

Lai J, Porreca F, Hunter JC, Gold MS. Voltage-gated sodium channels and hyperalgesia. *Annu Rev Pharmacol Toxicol* 2004;44:371–397.

Luo CH, Rudy Y. A dynamic model of the cardiac ventricular action potential: I. Simulations of ionic currents and concentration changes. *Circ Res* 1994;74:1071–1096.

Viswanathan PC, Balser JR. Inherited sodium channelopathies: A continuum of channel dysfunction. *Trends Cardiovasc Med* 2004;14:28–35.

Synapses 6

OBJECTIVES

- *Describe the steps in chemical synaptic transmission.*
- *Describe the biosynthesis and actions of acetylcholine.*
- *Describe the biosynthesis and actions of catecholamines (dopamine, norepinephrine, epinephrine).*
- *Describe the biosynthesis and actions of serotonin and histamine.*
- *Describe the biosynthesis and actions of excitatory and inhibitory amino acids.*
- *Describe the biosynthesis and actions of neuropeptides.*
- *Describe the structure of the neuromuscular junction and the functions of the various substructures.*
- *Describe and explain the steps involved in neuromuscular transmission.*
- *Describe the actions and explain the mechanisms for the effects of Ca and Mg on transmitter release.*
- *Describe how acetylcholine interacts with receptors on the postsynaptic membrane and the fate of the acetylcholine.*
- *Describe the generation of the endplate potential and the effects and mechanisms of action of acetylcholine esterase inhibitors and blockers of acetylcholine receptors.*
- *Describe facilitation and posttetanic potentiation of transmitter release and how these processes can be used to explain certain features of myasthenia gravis and recovery from receptor blockade.*
- *Describe the structures and explain the functions of the various parts of neurons.*
- *Describe transport of materials down axons (orthograde transport).*
- *Describe retrograde axonal transport, include mechanisms and materials.*
- *Discuss causes and consequences of disrupted axoplasmic transport.*
- *Calculate the time required for the regeneration of peripheral nerves.*
- *Describe the differences and similarities between synaptic transmission at a central synapse and at neuromuscular junctions.*
- *Describe the generation of IPSPs and EPSPs by ionotropic and metabotropic receptors.*
- *Describe the integration of information and repetitive firing in neurons and the concept of presynaptic inhibition.*

A synapse is a specialized region where a neuron communicates with a target cell: another neuron, a muscle cell, or a gland cell. The word was coined by Charles Sherrington more than a century ago; it is derived from Greek words meaning "to fasten together." Most synapses are chemical; the presynaptic neuron releases a transmitter substance that diffuses across the synaptic cleft and binds to a receptor on the postsynaptic cell. The postsynaptic receptor may be ionotropic, in which case it will open a selective pore and allow ions to flow to produce a postsynaptic potential, or it may be metabotropic and inform a G protein to initiate a chemical cascade, which may include the opening or closing of channels. A few synapses are electrical; current passes through cell-cell channels directly into the postsynaptic cell. Chemical synapses offer the possibility of amplification, signal inversion, and persistent effects; electrical synapses are faster and seem to be used when synchronization is more important than computation.

Chemical synapses may be excitatory or inhibitory, depending on their effect on the postsynaptic cell. In the CNS, neurons receive both types of synapses and integrate the information coming in to them before sending the processed message on to another cell. Chemical synapses are major pharmaceutical targets both medically and socially.

PRESYNAPTIC PROCESSES

The presynaptic terminal must provide for the synthesis, packaging, and release of the various transmitters (Fig. 6-1). The nonpeptide transmitters are concentrated inside the vesicle by specific H/transmitter cotransporters.

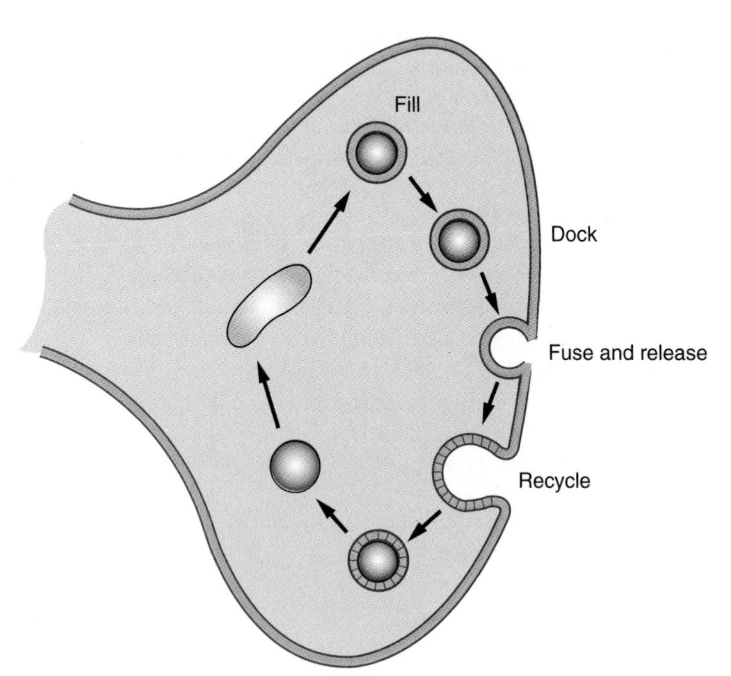

Figure 6-1. Synaptic vesicle docking, releasing of contents, and recycling.

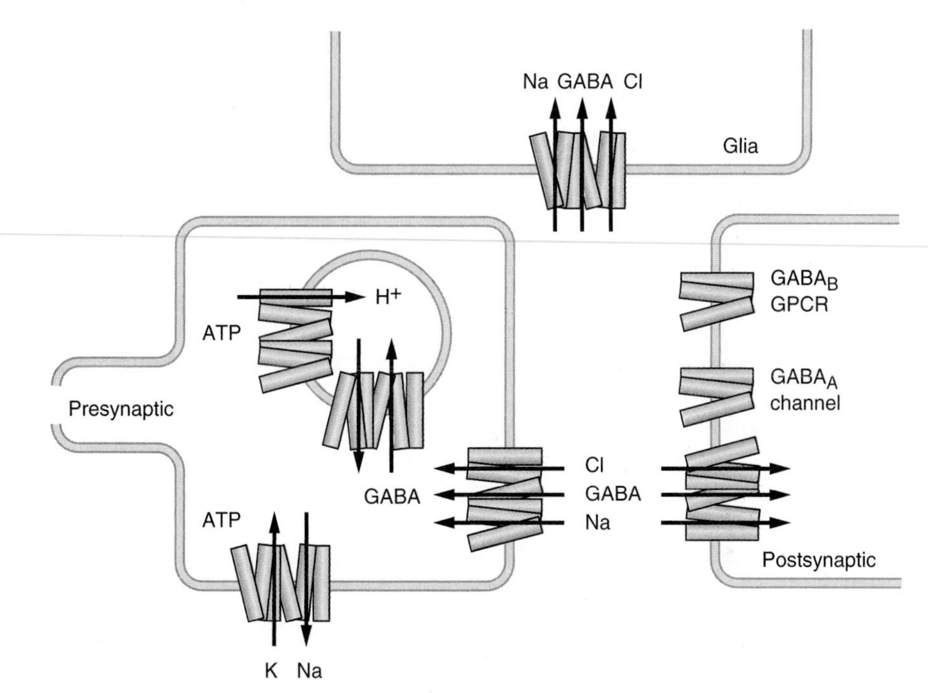

Figure 6-5. A generalized schematic GABA synapse.

biogenic amines are via GPCRs, often without producing postsynaptic potentials. All are concentrated into vesicles and released by similar mechanisms, but some are released by nerve swellings, which are in the vicinity of the receptors but not as closely apposed as in Fig. 6-19. Nonnerve cells also release EPI and histamine.

CATECHOLAMINES

Dopamine and NE are found in the CNS. NE is the principal final transmitter of the sympathetic nervous system and EPI is made and released by the adrenal medulla. All three are synthesized by the same pathway, starting with the hydroxylation of tyrosine to dihydroxyphenylalanine (**DOPA**), which is then decarboxylated to form dopamine. Adding a beta-hydroxyl group forms NE and, in the adrenal medullary cells, a subsequent transfer of an N-methyl group forms EPI. **Tyrosine hydroxylase** (TH) is the rate-limiting enzyme. TH and DOPA decarboxylase are in the presynaptic terminal cytoplasm. Dopamine is concentrated into vesicles, where dopamine beta hydroxylase (DBH) converts it into NE. NE is taken back into the presynaptic terminal by a Na/Cl coupled cotransporter; there, it is broken down by monoamine oxidase (**MAO**) in mitochondria and by catecholamine-O-methyl transferase (**COMT**) in the cytoplasm.

Catecholamine receptors are GPCRs and are found in the CNS, smooth muscles, and heart. **Adrenergic receptors** respond to NE and/or EPI. There are two categories of adrenergic receptors: **alpha-adrenergic receptors** have a higher affinity for NE and **beta-adrenergic receptors** have a higher affinity for EPI. However, there is cross-reactivity, and both receptors will respond to higher concentrations of both agonists. In the cardiovascular system, the alpha receptors are primarily found on the smooth muscle cells that control the diameter of small blood vessels; NE acts to constrict these vessels. The beta receptors are primarily in the heart and can make it beat faster and harder. Muscle relaxation via adrenergic receptor activation occurs in smooth muscle cells in the gut and the lungs. Some of these functions are discussed at greater length in Chap. 7.

Parkinson's disease is characterized by the loss of dopaminergic neurons; its treatment often includes DOPA, which can partially relieve the symptoms. Drugs that block dopamine receptors have been used to treat **schizophrenia**; sometimes they induce Parkinson-like tremors. **Reserpine**, an early tranquilizer, inhibits dopamine transport into vesicles. **Cocaine** blocks the reuptake of catecholamines, prolonging their actions. Many over-the-counter home remedies, such as **neosynephrine** and sudafed, activate catecholamine receptors.

Serotonin

Serotonin, or 5-hydroxytryptamine (**5HT**), is made from tryptophan by hydroxylation and decarboxylation. 5HT receptors function in the gut in secretion and peristalsis, mediate platelet aggregation and smooth muscle contraction, and are distributed throughout the limbic system of the brain. Serotonin was initially identified as a substance in blood serum that constricted blood vessels, hence the name.

Tryptophan hydroxylase is the rate-limiting step of 5HT synthesis; in the CNS tryptophan hydroxylase is present only in serotonergic neurons. 5HT is deactivated by reuptake and then broken down by MAO in mitochondria. Most 5HT receptors are GPCRs; $5HT_3$ receptors are ion channels.

Selective serotonin reuptake inhibitors such as fluoxetine hydrochloride (**Prozac**) are commonly prescribed as antidepressants. Lysergic acid diethylamide (**LSD**) and psilocin, the active metabolite of **psilocybin**, activate 5HT receptors.

Histamine

Histamine is released from mast cells (part of the immune system) in response to antigens or tissue injury. Histamine release is associated with allergic reactions; it initiates inflammatory responses, dilates blood vessels, decreases heart rate, and contracts smooth muscles in the lung. Enterochromaffin-like cells in the gastric mucosa also release histamine, which promotes acid production. Histamine is made from histidine, stored in vesicles, and released; it is then broken down by histamine N-methyl transferase.

PURINES

ATP is contained in synaptic vesicles and released with NE in sympathetic vasoconstrictor neurons. ATP induces constriction when applied directly to the smooth muscles. P2X ATP receptors are ion channels that permit the entry of Ca, and the cells also have P2Y GPCRs. These receptors are also in the brain, as well as P1 receptors for adenosine.

PEPTIDES

Neuropeptides are small polypeptides that are synthesized as larger inactive precursors (propeptides) and then cut out by specific endopeptidases. As they are proteins, they are synthesized in the cell body and transported in vesicles to the terminals. There is no reuptake mechanism. Peptides are less concentrated than other neurotransmitters in vesicles but have higher affinity for their receptors, which are GPCRs. Neuropeptides are released from large dense-core vesicles, while other neurotransmitters are secreted from smaller, clearer vesicles. Neuropeptides often act in concert with classic neurotransmitters.

Not much is known about the function of most neuropeptides in the CNS except the opiate peptides, **endorphin**, **enkephalin**, and **dynorphin**, which are involved in the regulation of pain perception. Three opiate receptors have been identified, initially as the sites that bind synthetic opiates such as **morphine**.

There are many nonopiod peptides released from neurons. The calcitonin gene–related peptide (**CGRP**) and **substance Y** are involved in maintaining blood pressure. Antidiuretic hormone (**ADH**, also called **vasopressin**) helps control water reuptake in the kidney. **Oxytocin**, luteinizing hormone (**LH**), and follicle-stimulating hormone (**FSH**) are involved in reproduction. Cholocystokinin (**CCK**), **gastrin**, and vasoactive intestinal peptide (**VIP**) facilitate digestion. All these and more have been identified as potential neurotransmitters in the CNS.

Synaptic Release

The details of the synaptic release process are currently under active investigation. It is clear that the process is triggered by an increase in cytoplasmic Ca levels. At many synapses, when a presynaptic action potential arrives, the Ca enters the terminal through Ca_V channels. In some small sensory cells there is no action potential and the sensory generator potential opens the Ca_V channels.

Synaptic vesicles cycle through loading with transmitters, docking at an active zone or release site, fusion with the surface membrane and release of contents, endocytotic recovery, and then loading again. In Fig. 6-6, each step in the vesicular cycle is illustrated by a shift in the position of the vesicle. In reality, however, there is little movement in the attached states. In many synapses, the release site is across from a postsynaptic area containing the channels that are sensitive to the transmitter. In the neuromuscular junction (see Fig. 6-9), Ca_V channels are adjacent to the release site, so that internal Ca need only be elevated locally to cause release.

Docking and fusion involves the **SNARE** or soluble N-ethylmaleimide–sensitive factor attachment protein (SNAP)-receptor proteins that are present on both membranes before fusion and associate into tight core complexes during fusion. Figure 6-7 shows the vesicular **v-SNARE synaptobrevin** binding the target **t-SNARE syntaxin** and **SNAP-25**. Synaptobrevin is the substrate of the endopeptidases contained in botulinum and tetanus toxins.

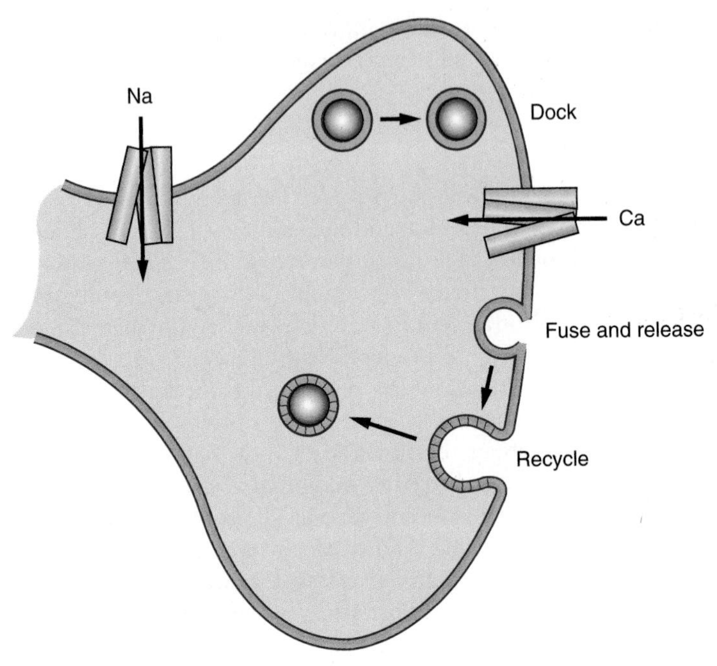

Figure 6-6. The channels involved in synaptic release.

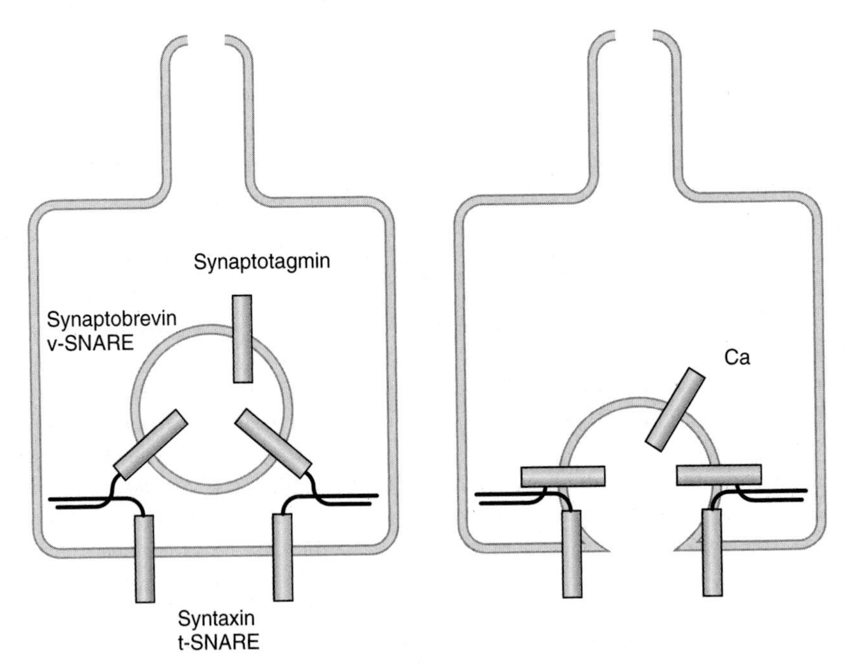

Figure 6-7. A possible mechanism of vesicle fusion.

Ca-stimulated fusion requires the Ca binding protein **synaptotagmin**, which is in the vesicular membrane and binds Ca. A proposed model suggests that Ca allows the synaptotagmin to bind the surface membrane and pull the two lipid layers together.

The recycling process returns the lipids and proteins to the vesicle pool. The vesicle is reformed as a **clathrin**-coated pit. The clathrin molecules have the shape of a triskelion, or three bent legs. The clathrin forms a closed surface covered with pentagons and pinches the recovered vesicle off the surface.

Axoplasmic Transport

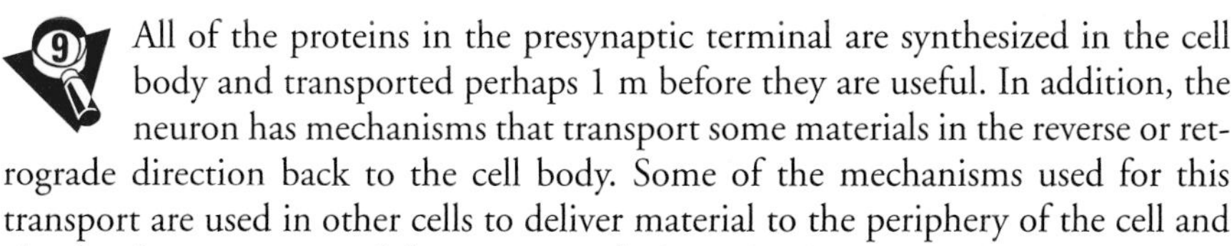

All of the proteins in the presynaptic terminal are synthesized in the cell body and transported perhaps 1 m before they are useful. In addition, the neuron has mechanisms that transport some materials in the reverse or retrograde direction back to the cell body. Some of the mechanisms used for this transport are used in other cells to deliver material to the periphery of the cell and also for the movement of chromosomes during mitosis.

Axoplasmic transport is distinguished by the direction into **orthograde** and **retrograde**. Orthograde transport can be further divided into fast (100 to 400 mm/day or 1 to 5 μm/s) and slow (0.5 to 4 mm/day). Fast transport is for vesicles and mitochondria; slow transport is for soluble enzymes and those that make up the cytoskeleton. Retrograde transport is only of the fast type.

Fast axoplasmic transport involves molecular motors that hydrolyze ATP and walk along **microtubules**, long hollow cylinders 25 nm in diameter. Two different classes of motors are used, **kinesins** for orthograde transport and **dyneins** for retrograde. Microtubules are polarized, and these motors can sense the polarity and move by 8-nm steps in the appropriate direction. The motors have two "feet," or sites of interaction with the microtubules, and exhibit **processivity**, or the ability to function repetitively without dissociating from their substrate, the microtubule. Accessory molecules are used to attach the payload to the motor (Fig. 6-8).

Fast orthograde transport delivers the membrane proteins needed in the terminal both for the vesicles and the terminal membrane. During development, it can also deliver cell adhesion molecules that recognize or induce targets. Retrograde transport can return damaged proteins for the endolytic pathway and bring information about signaling events back to the cell body.

Retrograde transport is part of the pathophysiology of several diseases including **polio, rabies, tetanus**, and **herpes simplex**. The herpesvirus enters peripheral nerve terminals and then travels back to the cell body, where it replicates or enters latency. It can later return to the nerve ending by orthograde transport and make itself available for contact transmission to another person. The tetanus toxin is transported retrogradely in motoneurons to the dendrites and then transsynaptically to GABA- and glycine-releasing terminals, where it inhibits synaptic release.

Axoplasmic transport is important for the regeneration of nerves following injury in the peripheral nervous system. Under usual circumstances, nerves in the CNS do not regenerate, although current researchers are hopeful that this will

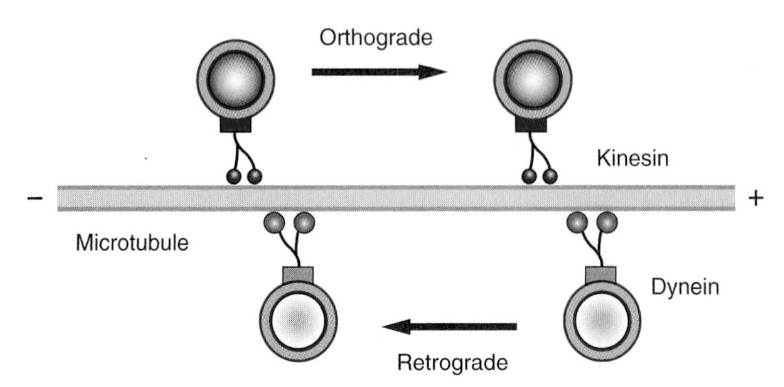

Figure 6-8. Axoplasmic transport.

change in the future. If a peripheral nerve axon is cut or crushed, the distal portion will die and go through a characteristic **Wallerian degeneration** as the axon is resorbed over a few weeks. Within a few days, the cell body undergoes the **axon reaction**, often called **chromatolysis**, because of a change in staining when it is studied histologically. The nucleolus enlarges, the rough endoplasmic reticulum, or ER (Nissl substance), disperses, and the nucleus is displaced. Genes have been activated, RNA transcribed, and proteins synthesized. The longer the distance from the injury to the cell body, the greater the latency, indicating that retrograde transport is involved in the signaling to initiate the axonal reaction.

At the site of injury, the end that is coupled to the cell body will reseal in hours and buds or sprouts will appear in a day or two. The cut tip swells with mitochondria and smooth ER. The sprouts grow out as thin fibers. If the regeneration is successful, one of the new fibers finds its way down the sheath of the distal degenerating nerve and reinnervates a postsynaptic target. The fiber will then increase in diameter and become remyelinated. The rate of fiber growth is about 1 mm/day, in the range of slow axonal transport. This is the number to use in estimating recovery times.

POSTSYNAPTIC PROCESSES

There are several different postsynaptic receptors for each transmitter; they are distinguished by their amino acid sequences and, in some cases, pharmacology. Different regions of the nervous system have characteristic receptors; sometimes an individual postsynaptic cell will have multiple receptor types. The ionotropic receptors are excitatory or inhibitory according to their ionic selectivity. The metabotropic receptors may indirectly cause channels to open or close and may also modulate the activity of the cells in other ways.

10 The postsynaptic potentials are called **excitatory postsynaptic potentials, EPSPs**, if their effect is to make the postsynaptic cell more likely to respond with an action potential, or **inhibitory postsynaptic potentials, IPSPs**, if they make the postsynaptic cell less likely to fire an action potential. Each channel has a selectivity pattern and allows different ions to flow through

with differing ease. This means that each channel will have a reversal potential or there will be some potential at which there will be no net flow of ions through the channel. If the membrane potential is more positive than the reversal potential, net current will flow out of the cell, tending to hyperpolarize it. If the membrane is less positive or more negative, current will flow in and tend to depolarize the cell. The current that flows through channels drives the membrane potential toward the reversal potential for that channel.

Most neurons in the CNS receive a constantly fluctuating input from a variety of synapses and their membrane potential is always changing. If a synapse opens channels having a reversal potential more positive than the threshold for action potentials, they will produce an EPSP. If the reversal potential is more negative than the threshold, an IPSP will result. If a channel is permeable to a single ion, its reversal potential is the Nernst potential for that ion (Eq. [3.4]). If the channel is permeable to multiple ions, its reversal potential is the weighted average of the Nernst potentials for its ions (Eq. [3.6]).

nAChR channels and GluR channels are approximately equally permeable to Na and K, and their reversal potential is about −10 mV; when activated, they make EPSPs. $GABA_A$R and glyR are Cl channels; their reversal potential is about −80 mV. The cardiac mAChR, through a G protein, activates a K_{ir} channel (K_{ACh}) that has a reversal potential about −90 mV. Both Cl channels and K channels make IPSPs. If for some reason the cell happens to be more negative than −80 mV, opening Cl channels will depolarize the cell but still work to keep other channels from further depolarizing the cell to threshold.

THE NEUROMUSCULAR JUNCTION—A SPECIALIZED SYNAPSE

Because of its easy accessibility, the neuromuscular (or myoneural) junction (Fig. 6-9) is the best-studied synapse; it is the source of much of what is known about synapses. This section describes the functioning of this synapse, bringing together and illustrating many of the ideas introduced more abstractly above. The neuromuscular junction is of considerable clinical interest. **Myasthenia gravis** is a disease that incapacitates the neuromuscular junction; there are other diseases and several drugs and toxins that target the junction. The neuromuscular junction provides a convenient assay for the anesthesiologist gauging recovery from muscle immobilization after surgery.

A single motoneuron controls between three and a thousand muscle cells. Each muscle cell receives input from one motoneuron. The combination of the motoneuron and all of its muscle cells function together as a **motor unit**. In healthy people, an action potential in the motoneuron will produce a large EPSP in all of its muscle cells, large enough to greatly exceed the threshold of the muscle cells and produce action potentials and contraction. The CNS regulates movement by choosing which motor units to activate. Smaller motor units produce finer movements.

At the nerve ending, the axon looses its myelin and spreads out to form the **motor endplate**, named for its anatomic appearance. The nerve terminals contain

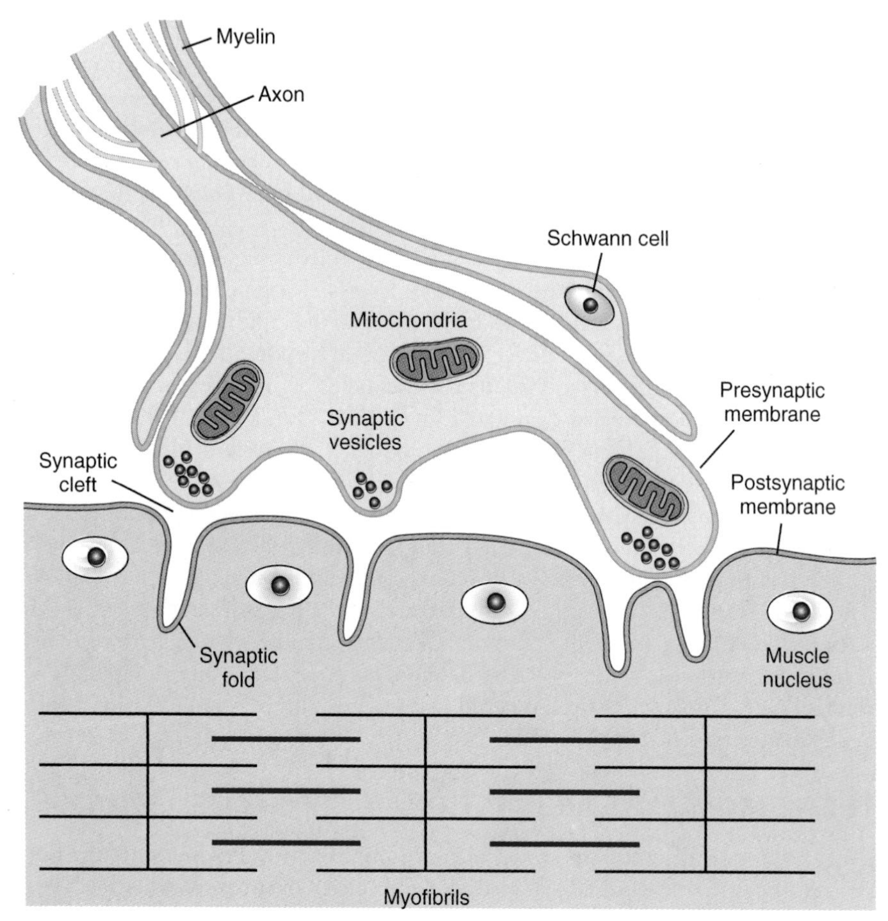

Figure 6-9. The neuromuscular junction.

many mitochondria and many 40-nm-diameter synaptic vesicles that contain acetylcholine (ACh). The nerve terminal is separated from the muscle by a 50-nm gap, the **synaptic cleft**, which contains a basal lamina. The muscle membrane contains ACh receptors (AChRs) and also acetylcholine esterase (AChE). In transmission electron micrographs, both pre- and postsynaptic membranes appear thickened, indicating the presence of channels and other proteins.

11 Neuromuscular transmission can be described as a 10-step process: (1) an action potential enters the presynaptic terminal; (2) the nerve terminal is depolarized; (3) depolarization opens Ca_V channels; (4) Ca enters the cell, moving down its electrochemical gradient; (5) Ca acts on a release site, probably synaptotagmin, causing synaptic vesicles to fuse with the presynaptic membrane; (6) about 200 vesicles release their ACh into the synaptic cleft; (7) the ACh in the cleft either (a) diffuses away out of the cleft, (b) is hydrolyzed by AChE into acetate and choline, or (c) interacts with AChRs on the postsynaptic membrane; (8) the activated

AChRs are very permeable to Na and K and slightly permeable to Ca, hence a net influx of positive charge into the muscle cell depolarizes the muscle membrane in the endplate region; (9) when the muscle membrane is depolarized to threshold, an action potential is elicited, which propagates in both directions to the ends of the muscle cell (the link between muscle excitation and contraction is discussed in Chap. 7); and finally (10) choline is recycled into the nerve terminal, Ca is pumped out of the nerve terminal, and vesicles are recycled and refilled.

Recording the Endplate Potential

If a microelectrode is inserted into a muscle fiber near the neuromuscular junction, a resting potential of about −90 mV will be measured. If the nerve is stimulated and the muscle is prevented from contracting by extreme stretching (see Fig. 7-4), the membrane potential will be seen to change, as shown in the solid trace on the left in Fig. 6-10. If, instead, the electrode is placed several centimeters away from the neuromuscular junction, the potential shown in the right trace will be seen. If the concentration of Ca in the bath is decreased, the concentration of Mg is increased, and the nerve is stimulated again, the potential at the neuromuscular junction will change, as shown in the dashed trace. Under these conditions there will be no change in the membrane potential several centimeters away from the junction.

The solid trace on the left shows an action potential superimposed on an endplate potential (EPP). There is an initial depolarization due to a net entry of positive charge through AChRs that were activated by the released ACh. When the potential reached about −50 mV, an action potential was initiated. In normal Ca, the endplate potential is two or three times larger than necessary to depolarize the muscle membrane to threshold.

The pure action potential is seen in the trace on the right; it can be recorded by stimulating one end of the muscle electrically or by placing the recording electrode a few centimeters away from the endplate. The dashed trace on the left shows an endplate potential with reduced amplitude. The endplate potential is not visible a few centimeters away from the endplate (right). A reduction in extracellular Ca

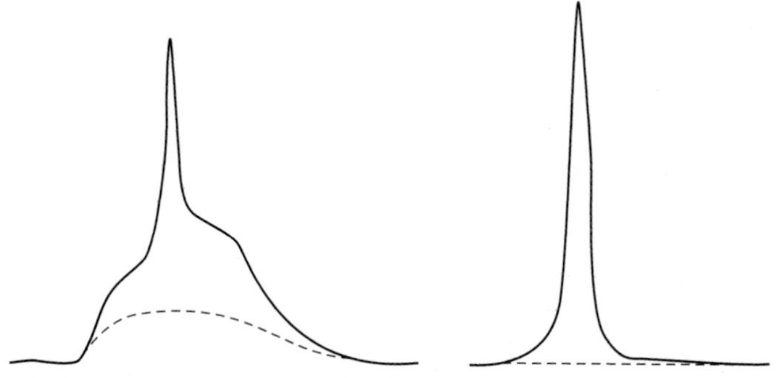

Figure 6-10. An endplate potential and action potential at the neuromuscular junction (left) and 2 cm away (right).

reduces the release of ACh and thus reduces the EPP. An increase in Mg reduces transmitter release by reducing Ca entry through Ca_V channels. These opposing effects of Ca and Mg have been seen on all chemical synapses that have been examined; this is now considered one of the tests for identifying a chemical synapse.

(12) Ca and Mg concentrations have a different effect on the excitability or the threshold for action potentials on the nerve and muscle cells. The reduction of Ca makes the cells more excitable or have a more negative threshold, or requires a smaller depolarization to reach threshold for an action potential. This is an effect on the Na_V channels; in low Ca, Na_V channels will open at more negative potentials. Ca and Mg have a synergistic action on Na_V channels; they have opposing actions on neuromuscular transmission. Clinically, the effects of hypocalcemia are hyperexcitability and spontaneous action potentials in nerve and muscle. These effects are seen when there is still enough Ca to support sufficient ACh release, so that every nerve action potential leads to a muscle action potential.

In this low-Ca and high-Mg case, the EPP is not large enough to reach threshold and elicit an action potential. The action potential is actively propagated; the endplate potential spreads passively and will not be visible a few centimeters from the neuromuscular junction. These two potentials are produced by the activity of different channels that have differing pharmacology. Curare will block AChRs and the endplate potential without affecting the action potential seen following direct electrical stimulation of the muscle. A toxin from a cone snail (μ-conotoxin) will block the muscle action potential but not the endplate potential. The μ-conotoxin blocks muscle Na_V channels but not nerve Na_V channels, which are a different gene product.

If the Ca/Mg ratio is sufficiently low, the response to stimulation will appear as in Fig. 6-11. Each trace represents the response to a stimulation that is repeated every 5 s. Three of the traces show a small endplate potential; in the third trial there was no response. The first response is about 1 mV high; the second and fourth responses are about 0.5 mV. When this experiment was repeated many times, the

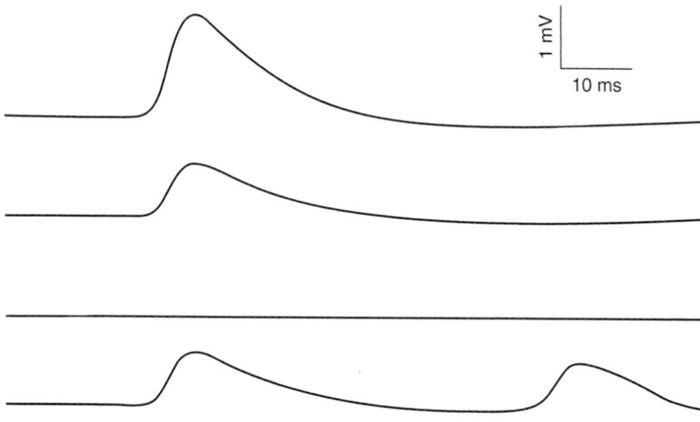

Figure 6-11. Some miniature endplate potentials.

responses were found to be quantized with a unit response of about 0.5 mV. That is, there were many 0.5-mV, 1-mV, and 1.5-mV responses but very few with amplitudes in between. In addition, there are sometimes spontaneous 0.5-mV responses without any stimulation; one of these was caught on the fourth trace. These miniature endplate potentials (MEPPs) represent the postsynaptic response to the release of 1, 2, or 3 **quanta** of ACh. Each quantum is the contents of a single vesicle. The exact number of vesicles released on any particular stimulation cannot be known; only the average number or the **mean quantal content** can be predicted. The EPP in normal Ca/Mg conditions is the response to about 200 quanta.

The average rate of spontaneous MEPPs is about 1 vesicle per second. In a normal endplate potential, the 200 vesicles are released within 1 ms, which means that stimulation increased the rate of release by 200,000-fold. If black- (or brown-) widow spider venom (BWSV) is applied to a neuromuscular junction, the MEPP frequency increases to a few hundred per second for about 30 min and then stops. In total about 200,000 vesicles are released, which is equal to the number seen by the electron microscope in an unstimulated neuromuscular junction. After BWSV treatment, no vesicles are visible. BWSV paralyzes by depleting the nerve terminals of synaptic vesicles. It can be deadly if the nerve endings controlling breathing are compromised.

Transmitter-Receptor Interaction

The nicotinic ACh receptor at the neuromuscular junction has five subunits, each with four TM segments. Two of the subunits are called alpha subunits and bind ACh at the α–γ and α–δ interfaces near the top of the molecule, about 5 nm from the center of the membrane. The channel then undergoes a conformational change that is transmitted through the molecule to open the pore, most likely by causing the M2 TM segments to move out away from the axis of the pore, making it larger. The open pore allows Na and K and, to a lesser extent, Ca to pass. The pore stays open about 1 ms and about 20,000 ions pass at a rate of 2×10^7/s, which is equivalent to about 3 pA. If a single AChR is captured in a patch of membrane and maintained with a –90 mV potential, application of ACh will cause the channel to open and close several times, each opening appearing as a 3-pA current pulse of varying duration averaging about 1 ms. A single quantum opens about 2000 channels; 200 quanta open about 400,000. A neuromuscular junction has many more channels, about 20 million; thus, only a small fraction is used at any one time.

13 The number of open channels is proportional to the concentration of ACh squared and the effective number of receptors. A kinetic scheme for the reaction is shown in Fig. 6-12. The receptor can open with one or two ACh molecules bound; it stays open about 10 times longer with two bound. It is the concentration of R*2ACh that is proportional to the concentration of ACh squared.

$$\text{Number of open channels} = k[R][ACh]^2 \qquad [6.1]$$

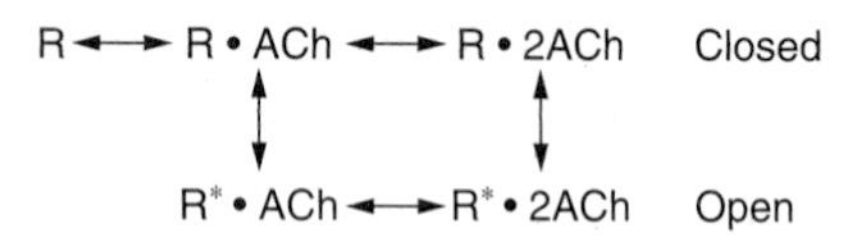

Figure 6-12. A kinetic scheme of the reaction between acetylcholine and the nicotinic acetylcholine receptor.

Desensitization

If a single AChR is exposed to continuous ACh for several minutes, its response will slow and openings will become less frequent. If ACh is added to the bath containing a neuromuscular junction, the muscle membrane potential will depolarize but the response will reach a peak and then decline, as shown in Fig. 6-13. This decline is called **desensitization**; the AChR molecule has entered an inactivated state from which it does not open. This is functionally somewhat similar to the inactivation of Na_V channels except that the time course, the agent that causes the inactivation, and the molecular basis in the channels are completely different. Desensitization probably does not occur with normal use of neuromuscular junctions but may become a problem when drugs are used that block acetylcholine esterase. A patient with desensitized AChRs may be paralyzed and unable to breathe due to a lack of functional AChRs.

Some Drugs That Act at the Neuromuscular Junction

(14) D-Tubocurare is a classic neuromuscular blocking agent, originally discovered as an arrow poison from South America. Curare binds AChRs reversibly and prevents ACh from opening the channels. After application of curare, the EPP becomes smaller; if there is sufficient curare, the EPP becomes so small that it no longer elicits an action potential, similar to the dashed response in Fig. 6-10, and the junction is effectively blocked. Higher doses of curare can eliminate the EPP. Curare reduces the EPP by reducing the number of receptors available to respond to ACh. Curare or a related drug is often used during surgery to immobilize muscles; it can also facilitate tracheal intubation or mechanical ventilation.

Anticholinesterases such as **neostigmine** combine with AChE and prevent hydrolysis of ACh, which leads to a larger EPP. Neostigmine is used to speed recovery from the effects of curare and to reduce the symptoms of myasthenia gravis (see below). There are dangers associated with neostigmine; an excess of ACh can lead

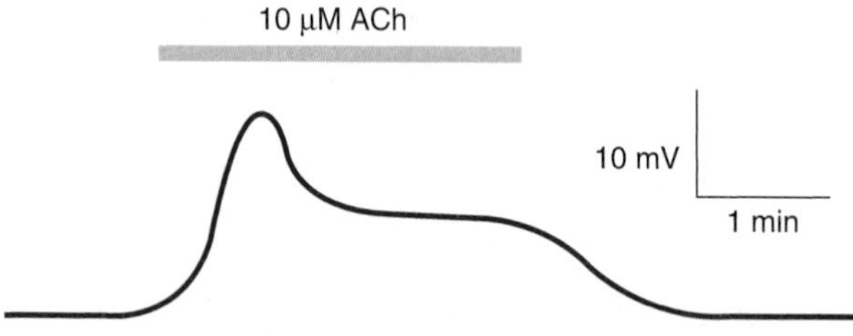

Figure 6-13. Desensitization of acetylcholine receptors.

to desensitization of the remaining receptors. Also, the body uses ACh to slow the heart and release saliva; both of these effects may be enhanced by physostigmine.

Botulism is a potentially fatal food poisoning caused by the anaerobic bacterium *Clostridium botulinum.* Some of the toxins released by this organism are endopeptidases, which are taken up by nerve cells and cleave synaptobrevin, thus preventing transmitter release. The purified toxins are used clinically to prevent unwanted neuromuscular transmission.

The *Bungarus* snake paralyzes its prey with **α-bungarotoxin**, which binds AChRs irreversibly and prevents their opening. Bungarotoxin has been fluorescently labeled and used experimentally to identify and locate nAChRs.

Myasthenia Gravis

Myasthenia gravis is a disease associated with muscle weakness and fatigability on exertion. It is an autoimmune disease that leads to the destruction of AChRs. Patients may have only 10 to 30 percent of the normal number of AChRs. Treatment with anticholinesterases increases the amount of available ACh, which makes it more likely that the remaining AChRs will be activated (Eq. [6.1]). There is a danger of giving too much anticholinesterase, which can lead to desensitization of the AChRs and further weakness. If this weakness is misinterpreted as insufficient anticholinesterase therapy, a tragic positive feedback loop leading to myasthenic crisis can ensue.

Lambert-Eaton Syndrome

The Lambert-Eaton syndrome is seen with an autoimmune disease that reduces the number of Ca_V channels in the presynaptic terminal. With prolonged effort, these patients gain strength, the opposite of myasthenic patients. Prolonging the presynaptic action potentials with drugs that block K_V channels, such as diaminopyridine, may alleviate some of the symptoms. The prolonged depolarization opens the remaining Ca_V channels for a longer time, allowing more Ca entry and therefore more release. If the experiment shown in Fig. 6-11 is performed on these neuromuscular junctions, they will be found to have a lower quantal content; that is, they release a lower number of vesicles per stimulus. This is in contrast to myasthenia gravis, which will show the normal quantal content but a smaller MEPP, the depolarization for each quantum.

Repetitive Stimulation

15 The amount of transmitter released by a synapse is not constant from impulse to impulse but depends on the past history of activity. If the nerve leading to a neuromuscular junction is stimulated once every 10 s or slower, it will consistently release about 200 vesicles. If the stimulation rate is abruptly changed to 50/s, which is roughly the rate used by the CNS to cause normal muscle contraction, the amount released per impulse will increase in the first half second and then decrease (Fig. 6-14). The increase, called **facilitation**, is

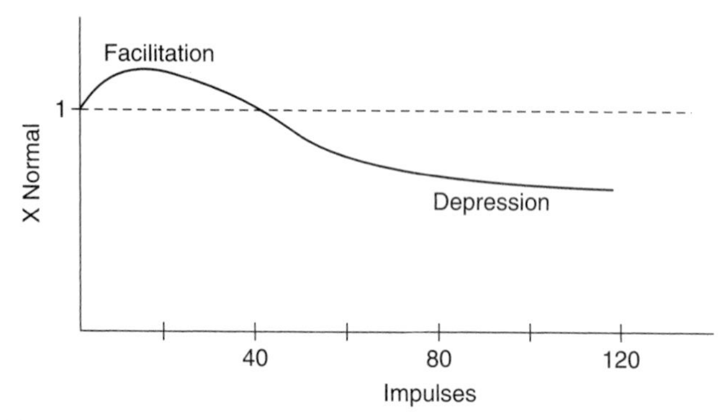

Figure 6-14. Facilitation and depression of synaptic transmission at the neuromuscular junction.

related to a buildup in residual calcium in the nerve terminal. The decrease, called **depression**, is thought to reflect a depletion of vesicles at the release sites.

This variation does not affect the functioning of a normal neuromuscular junction. Each of those nerve impulses releases sufficient ACh to produce an EPP large enough to fire a muscle action potential. However, the myasthenic person may have functional neuromuscular transmission only early in the task and experience weakness as the depression occurs with a prolonged effort and the amount of ACh released falls below what is necessary to trigger a muscle action potential. An anticholinesterase with a short duration of action, edrophonium chloride (**Tensilon**) is often used as a test for myasthenia gravis in patients who show rapid weakening when asked to perform a sustained contraction.

Posttetanic Potentiation

When the 50/s stimulus is stopped, there is an increase in the amount of transmitter that can be released by a single nerve impulse (Fig. 6-15). The nerve was stimulated once every 30 s before and after the tetanic stimulation. During the tetanus, the release increased and decreased, as in Fig. 6-14. After the tetanus, as the synapse recovered from depression, a posttetanic potentiation (**PTP**) was seen that lasted for several minutes. PTP is also related to an increase in residual Ca concentration in the nerve terminal, but PTP has a slower onset and slower decline than facilitation.

PTP is used as a diagnostic procedure following surgical procedures when curare or other neuromuscular blocking agents have been used to prevent unwanted motion. The anesthesiologist will give the patient anticholinesterase inhibitors, but she or he wants to know when just enough inhibitor has been given to avoid giving too much and desensitizing the AChRs. The anesthesiologist will repeat the experiment shown in Fig. 6-15, stimulating the thenar branch of the patient's median nerve and feeling the strength of contraction of the thenar muscles. Two shocks are

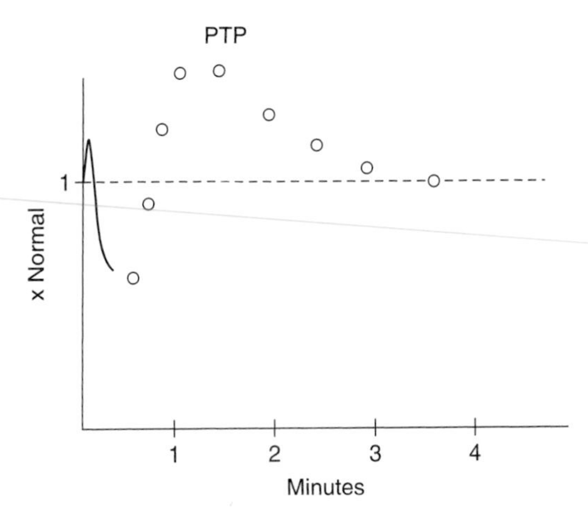

Figure 6-15. Posttetanic potentiation (PTP) of synaptic transmission at the neuromuscular junction.

given before the tetanus and then one 30 s later. Under deep curare, none of these will produce a palpable contraction. As more ACh is made available by blocking the esterase, the stimulus following the tetanus will give a larger response than the two before the tetanus because it will be the first one with an EPP large enough to excite the muscle. The endpoint is when enough esterase has been given that all three responses are the same because all three EPPs are above threshold for muscle activation.

Autonomic Synapses

The autonomic nervous system (ANS) has two divisions, both with two synapses outside the CNS (Fig. 6-16). The synapse closer to the CNS is referred to as the **ganglionic** synapse; the nerves leading into and out of the ganglia are called **preganglionic** and **postganglionic**. The sympathetic ganglia lie in a chain adjacent to the spinal column; the parasympathetic ganglia are close to the end organs where the second synapse occurs. The second synapses are onto smooth muscles or cardiac cells or gland cells. Many tissues receive both sympathetic and parasympathetic innervation.

The primary transmitter in the ganglionic synapse of both divisions is ACh; the receptors are nicotinic nAChRs that are heteromeric pentamers of related but different gene products than the nAChR of the skeletal muscle. The ganglionic receptors are less sensitive to curare and more easily blocked by hexamethonium. The primary postganglionic transmitter in the sympathetic nervous system is norepinephrine (NE), and there are two categories of GPCRs on the postsynaptic cells called alpha- and beta-adrenergic receptors. The primary postganglionic receptor in the parasympathetic division is ACh and the receptors are muscarinic mAChRs, which are also GPCRs but generally with different G proteins than the NE receptors.

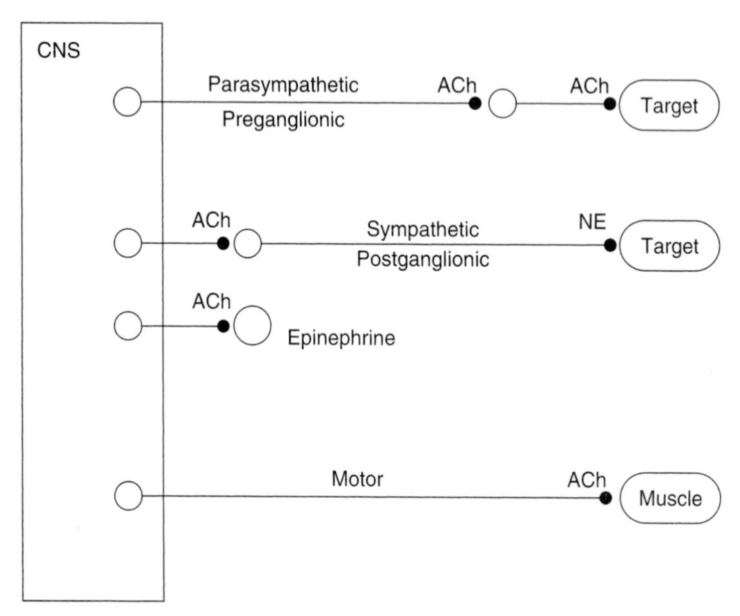

Figure 6-16. A schematic view of the efferent fibers of the autonomic nervous system and the motoneurons.

The ganglionic synapses are usually described as behaving more or less like the neuromuscular junction. However, the situation is more complicated; the postsynaptic neurons have dendrites with more than one presynaptic nerve ending on them. Different subpopulations of pre- and postsynaptic cells have been distinguished by looking at the peptide transmitters that are in these cells along with their classic transmitters. Postganglionic cells also have mAChRs that produce a slow EPSP by closing a K channel. There are also small, intensely fluorescent (SIF) cells in the ganglia that are innervated by preganglionic fibers and release NE or dopamine. All in all, it seems that some computation must be carried out in the ganglia, more than the simple pass-through circuit seen at the neuromuscular junction.

The synapses between the postganglionic cells and the end organs are different than the neuromuscular junction. The presynaptic processes are similar, but postganglionic cells make "**en passant**" synapses. Vesicles are seen in varicosities of the presynaptic nerve, which continues on to other varicosities before reaching its terminal.

Activation of mAChRs by the ANS increases GI tone and motility, increases urinary bladder tone and motility, increases salivation and sweating, and decreases heart rate and blood pressure. The ANS activates α-, β_1-, and β_2-adrenergic receptors, and α-adrenergic receptors raise blood pressure. β_1-adrenergic receptors increase heart rate and strength of contraction and blood pressure. β_2-adrenergic receptors dilate bronchioles in the lungs. The mechanism of the effects on cardiac and smooth muscle is discussed in Chap. 7.

Many agonist and antagonist drugs have been used to control these processes, some with more specificity than others. Thus there are specific α agonists and β blockers. Amphetamines and cocaine have an indirect adrenergic effect by stimulating NE release. Some compounds, like ephedrine, have both direct and indirect adrenergic effects. Atropine is the archetypical mAChR antagonist; its effects are the opposite of those attributed to ACh above. At many sites there is a tonic release of both ACh and NE from the ANS, so the blocking of one set of receptors may produce effects similar to activating the other.

CENTRAL NERVOUS SYSTEM SYNAPSES

(17) The human CNS has billions of neurons with trillions of synapses between them. A single neuron may have thousands of both excitatory and inhibitory inputs; some larger neurons may have over 100,000 endings on them. In order to accommodate this **convergence** of synaptic inputs, most neurons have a dendritic tree that greatly expands the area available for synaptic contact. The cell body (**soma**) and the initial region of the axon (**axon hillock**) integrate the incoming synaptic signals and determine when and how often the neuron will fire action potentials (Fig. 6-17). The axon carries the output of the neuron to the next group of neurons or to skeletal muscle cells if it is a motoneuron. Usually only a single axon leaves the cell body, but it later branches to allow

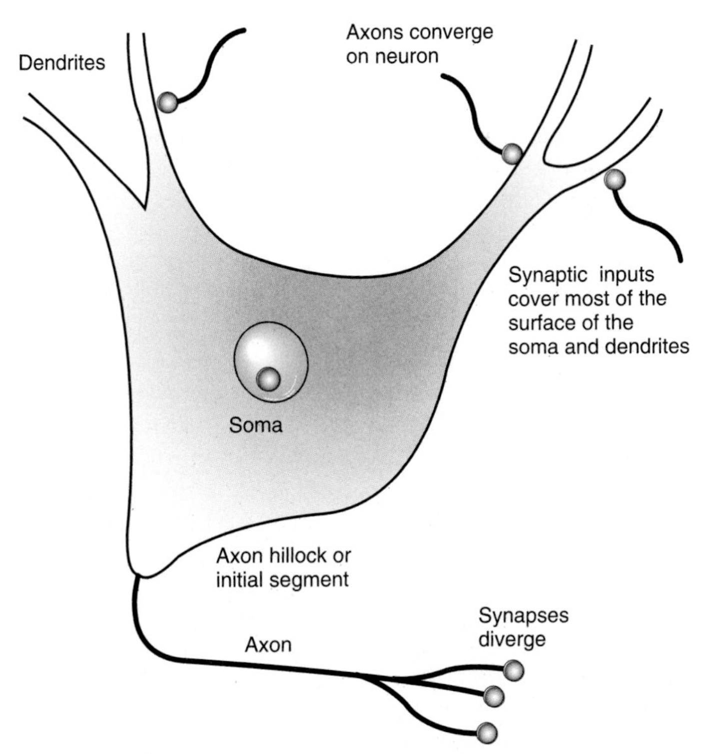

Figure 6-17. The convergence and divergence of synapses in the CNS.

the neuron to synapse with many other cells. This **divergence** of information combined with the convergence of many inputs onto a neuron gives the CNS much of its computational power.

Each neuron in the CNS acts as one or more small computers. While each cell performs its computations in milliseconds, millions of times slower than the central processing unit of a modern computer, the billions of neurons operating in parallel make the CNS shine in comparison. The CNS is capable of creating every thought in recorded history while simultaneously regulating both walking and chewing gum. The synapses make this possible. Learning and memory are accomplished by the modification of synapses.

There are two general types of synapses in the CNS, electrical and chemical. **Electrical synapses** operate by direct electrical current flow from the presynaptic neuron into the postsynaptic neuron through gap-junction channels between the membranes of the two cells (Fig. 6-18). Neurotransmitters are not involved and electrical synapses can have less synaptic delay than chemical synapses. However, unlike chemical synapses, electrical synapses cannot amplify the signal, nor can they reverse the direction of current flow. Gap junctions, which work as electrical synapses and allow action potentials to flow selectively from one cell to another, also connect cells in the heart and some types of smooth muscle.

There are two general types of chemical synapses in the CNS, excitatory and inhibitory. Excitatory synapses generate EPSPs that depolarize the membrane toward threshold. Inhibitory synapses generate IPSPs that either hyperpolarize the membrane or resist depolarization to threshold. Each of these types can be further divided into chemosensitive ion channels (or ionotropic receptors) and G-protein–linked ion channels (or metabotropic receptors). Chemosensitive ion channels typically give rise to fast synaptic events that last a few milliseconds; G-protein–linked ion channels may produce effects for hundreds of milliseconds.

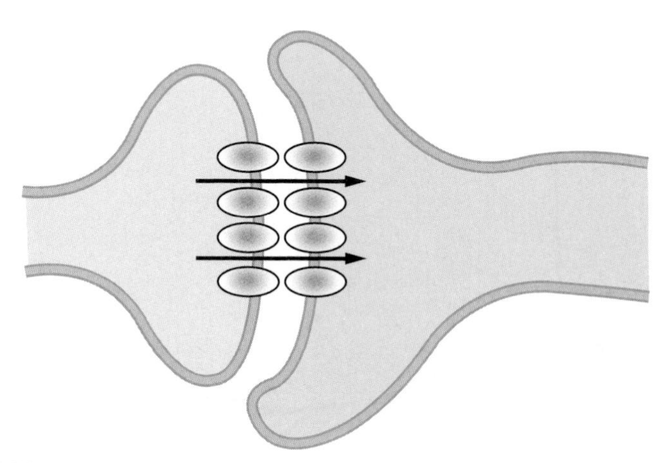

Figure 6-18. An electrical synapse.

INTEGRATION OF SYNAPTIC CURRENTS

Excitatory and inhibitory synapses inject current (positive or negative) into cells. These currents flow into the cell body and are summed. The PSPs passively spread to the spike initiation site or the part of the cell with the lowest threshold because of the cable properties of the cell. More distal synapses will be decremented compared to those near the site. The cell produces the spike initiation site by controlling the local density of Na_V channels. Often the spike initiation site is the axon hillock near the start of the axon (Fig. 6-17) or at the first node of Ranvier.

Because the PSPs last for several to many milliseconds, they can add together even though they do not occur synchronously; this is called **temporal summation**. The effects of synapses at different locations on the same postsynaptic cell can also add up; this is called **spatial summation**. Spatial summation is weighted inversely by the distance from the synapse to the initiation site of the action potential.

Figure 6-19 is a schematic drawing of a chemical synapse in the CNS. The presynaptic terminal is about 1 μm in diameter and contains mitochondria and synaptic vesicles filled with neurotransmitter. Depolarization of the terminal opens Ca_V channels, and Ca flows down its electrochemical gradient to act on synaptotagmin and trigger the fusion of a few vesicles with the presynaptic membrane in order to exocytose the neurotransmitter. The membrane is then recycled and the vesicles are refilled. The postsynaptic receptors are often on protrusions from dendrites, called spines, although synapses are also found on the dendritic shaft, the neuronal cell body, and other synaptic endings.

CNS synapses share many features with the neuromuscular junction but differ in several important respects. CNS synapses are much smaller and release far fewer vesicles, typically less than 5 per impulse, compared to about 200 at the motor endplate. In the CNS, synaptic clefts are narrower, about 20 nm, and cadherins and other cell adhesion molecules span the gap. ACh is the transmitter at the neuromuscular junction; there are a wide variety of transmitters in the CNS. The endplate

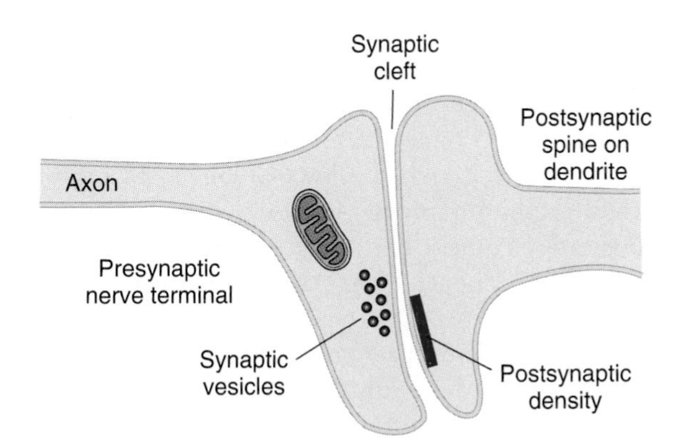

Figure 6-19. A CNS synapse.

potential is always excitatory and large enough to bring the muscle membrane to threshold; synapses in the CNS are excitatory or inhibitory and threshold is reached by the combination of hundreds of epsps.

There are some exceptional CNS synapses. In the cerebellum, a climbing fiber axon may make dozens of synapses on a Purkinje cell. In the calyx of Held, in the auditory pathway, the presynaptic ending forms a cap with finger-like stalks that envelop the postsynaptic neuron, covering about 40 percent of its soma. At both of these synapses a single presynaptic impulse releases hundreds of quanta, and the resulting EPSP is large enough to trigger a postsynaptic action potential.

Glutamate is the principle excitatory neurotransmitter in the CNS. There are several postsynaptic glutamate receptors, both channels and GPCRs. The channels can be grouped into two major types, **NMDA** and **non-NMDA** channels, according to their sensitivity to the synthetic agonist N-methyl-D-aspartate. Both types respond to glutamate. The non-NMDA channels may be called **AMPA, quisqualate**, or **kainate** channels, according to which of these nonphysiological agonists opens them. Non-NMDA channels typically generate fast EPSPs lasting about 5 ms.

When they are activated by glutamate, non-NMDA channels allow Na and K to flow through their pores. Each ion moves in the direction that will tend to bring the membrane potential to its Nernst equilibrium potential. Because both are moving, the membrane potential tends to approach the average of the two equilibrium potentials, which is about −10 mV. This potential, where the two ionic currents are equal, is called the reversal potential for the channel. When these channels open at potentials more negative than the reversal potential, the tendency for Na to enter the cell will dominate and the membrane will depolarize toward the reversal potential. If the starting potential were more positive than the reversal potential, the K ions would dominate and the cell would hyperpolarize toward the reversal potential.

NMDA receptor channels generate EPSPs lasting hundreds of milliseconds. Open NMDA channels allow Na and K and also Ca to pass through their pores. In the presence of glutamate, NMDA channels open only if the postsynaptic cell is also depolarized by some other means. This dual control of Ca entry has a key role in learning, as discussed below.

GABA (gamma-aminobutyric acid) is the major inhibitory transmitter in the brain. **Glycine** is an inhibitory transmitter in the brainstem and spinal cord. GABA opens **$GABA_A$** channels directly, which allows Cl ions to pass through their pores. GABA can also cause inhibition through **$GABA_B$** receptors, which are GPCRs that lead to the opening of K channels. The reversal potential for $GABA_A$ channels is at the Nernst potential for Cl, about −80 mV. If the membrane is more positive than E_{Cl}, Cl will enter the cell and make the membrane potential more negative, which will make it less likely to initiate an action potential.

Benzodiazepines such as diazepam (**Valium**) and **barbiturates** enhance the open probability of activated $GABA_A$Rs. Both have been used as sedatives and anticonvulsants. **General anesthetics** such as ether, chloroform, and halothane increase the duration of IPSPs and decrease the amplitude and duration of EPSPs.

CNS-MODULATORY NEUROTRANSMITTERS

In the CNS, ACh, NE, dopamine, and serotonin primarily act as diffuse modulators of activity as opposed to being involved in specific discrete tasks. Each of these neurotransmitters has its own set of neurons and targets; some of these neurons may contact more than 100,000 postsynaptic neurons. The postsynaptic receptors are metabotropic and alter the responsiveness of the postsynaptic neurons through second-messenger pathways. There are also ionotropic nAChRs in the CNS, but there are 10 to 100 times more mAChRs. The ACh and NE modulatory systems are part of the ascending reticular activating system that arouses the forebrain in response to stimuli. In some general ways, the modulatory systems play a role in the CNS similar to the role played by the ANS in the rest of the body.

PRESYNAPTIC INHIBITION

Some CNS synapses act directly on other synaptic endings rather than on dendrites or cell bodies (Fig. 6-20). Terminal A releases GABA on to terminal B, activating Cl channels that tend to hyperpolarize terminal B. If an action potential arrives in B while the Cl channels are open, the action potential's amplitude will be reduced, so that it will open fewer Ca_V channels and therefore fewer vesicles will be released by terminal B, and it will have a smaller effect on neuron C.

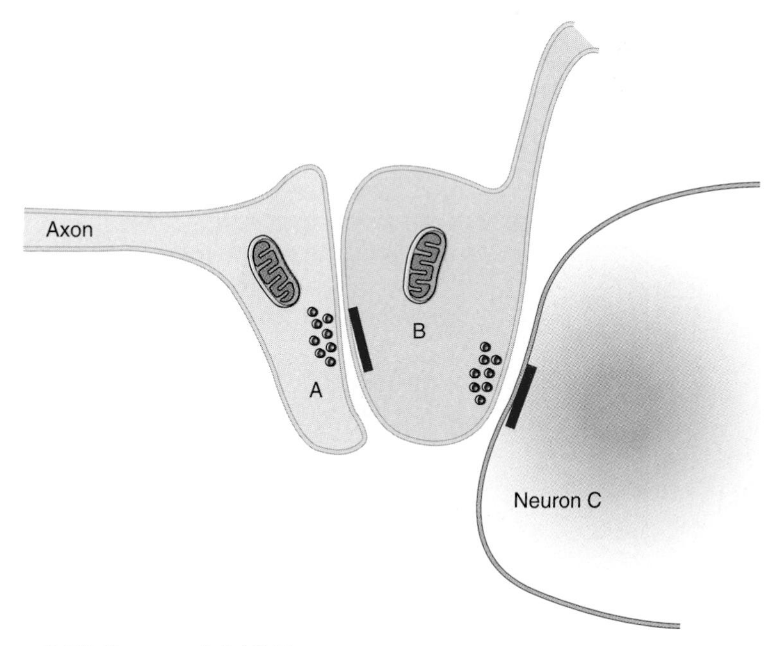

Figure 6-20. Presynaptic inhibition.

Retrograde Freely Diffusible Chemical Transmitters

In addition to the classic transmitters that are released from vesicles and bind to receptors, there are chemical messengers in the CNS with a different mode of operation. **Nitric oxide** (**NO**) is not stored but rather produced when needed. NO can freely diffuse across cell membranes from the inside of one cell (typically a postsynaptic cell body) to the inside of other cells (typically presynaptic endings), where it alters some chemical reactions. NO may spread to several presynaptic endings in the vicinity. NO is removed from the tissue by binding to hemoglobin.

Anandamide, an endogenous cannabinoid, is also produced as needed in postsynaptic cells and reaches the extracellular space by a nonvesicular process. It binds to presynaptic **cannabinoid receptors** (**CB1**), which are GPCRs, and can alter the subsequent release of traditional neurotransmitters.

Repetitive Firing of Nerve Cells

19

If an axon or a muscle cell is subjected to a maintained depolarization, it will respond with one or perhaps two action potentials and then stop firing because the Na_V channels enter the inactivated state and require a sojourn near the resting potential to recover. Many CNS cells and the slowly adapting sensory nerve endings will respond to a sustained depolarization with a train of action potentials at about 50/s. This is made possible by Ca_V channels and Ca-activated K channels. The action potential depolarization opens the Ca_V channels and the Ca that enters opens the Ca-activated K channels by binding to the intracellular portion of the molecule. The Ca-activated K channel then allows K to leave and the membrane potential to approach E_K for a long-lasting hyperpolarization, long enough for the Na_V channels to recover from inactivation (Fig. 6-21). The balance between the sustained stimulus and the rate that Ca is removed from the Ca-activated K channels determines the firing rate.

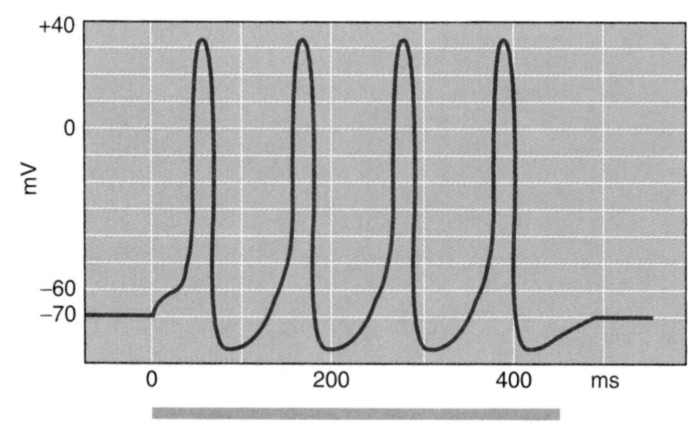

Figure 6-21. Repetitive firing of a motoneuron.

Learning, Memory, and Synaptic Plasticity

The cellular basis of learning and memory is a functional remodeling of synaptic connections, often called synaptic plasticity. This includes both **explicit** or **declarative** memory when the person can recall and describe some fact or past event and **implicit** or **procedural** memory, as in a learned motor skill. Memory is often subdivided into **short-term**, minutes to hours, and **long-term**, days to a lifetime. Short-term memory formation involves the modification of existing proteins, often by phosphorylation. Long-term changes involve gene activation, protein synthesis, and membrane rearrangement, including the formation and/or resorption of presynaptic terminals and postsynaptic spines. In a few studies, the volume of cerebral cortex dedicated to a task has been shown to increase with specific training.

The most intensively studied cellular learning phenomenon is **long-term potentiation** (**LTP**) in hippocampal synapses. The **hippocampus** is required for the formation of new long-term memories. If both hippocampi are compromised, the person will live continuously in the present with no recollection of events after the damage. In the hippocampus, LTP occurs at glutamate synapses between presynaptic CA3 cells and postsynaptic CA1 cells. LTP and the related **long-term depression** (**LTD**) also occur in other locations in the CNS. The classic experiment is similar to the PTP demonstration shown in Fig. 6-15; the synapse is tested infrequently, subjected to high-frequency stimulation, and then tested infrequently again. Unlike PTP, which disappears in a few minutes, with LTP, the potentiation remains for many hours or days (Fig. 6-22).

Also unlike PTP, LTP is primarily a postsynaptic event. It is not necessary to provide the high-frequency stimulation to the presynaptic terminals; simple depolarization of the postsynaptic cell paired with the presynaptic stimulation will induce LTP. This response to pairs of inputs made LTP a candidate basis for **associative** learning. There are two types of glutamate receptors on the postsynaptic membranes: AMPA (non-NMDA) and NMDA receptors. During low-frequency

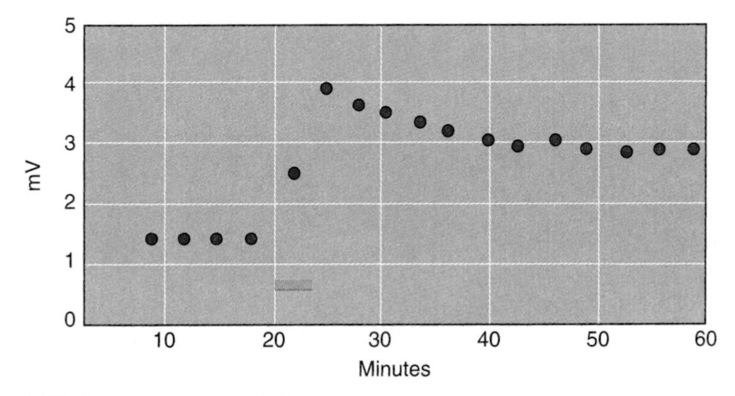

Figure 6-22. Long-term potentiation.

unpaired stimulation, only the AMPA receptors are activated; the NMDA receptors are plugged by external Mg ions. The AMPA receptor channels are permeable to Na and K; near the resting potential, Na movement into the cell is favored. When the postsynaptic membrane is depolarized, either by high-frequency synaptic input or by injecting current into the postsynaptic cell, the Mg is driven off the NMDA receptors and they respond to glutamate and allow Na and Ca to enter the cell. The elevated Ca activates a series of biochemical events that lead to the insertion of more AMPA receptors into the postsynaptic membrane.

LTP has been associated with learning in rats using a water maze. Rats with surgically removed hippocampi do not learn the maze. Neither do rats that have been treated with a specific antagonist for NMDA receptor channels. There are other examples of synaptic plasticity in other regions of the brain and there may well be additional mechanisms, including retrograde action of NO or anandamide.

KEY CONCEPTS

Synapses may be chemical or electrical. Chemical synapses may be excitatory or inhibitory.

In chemical synapses, the presynaptic terminal packages a neurotransmitter into vesicles. When the synapse is activated, the vesicle's contents are released and then a recycling process recovers some of the released transmitter and vesicular components.

Acetylcholine is the neurotransmitter at the neuromuscular junction. It is also an important component of the autonomic and central nervous system synapses.

Glutamate is the major excitatory neurotransmitter in the CNS.

GABA and glycine are the major inhibitory neurotransmitters in the CNS.

Several biogenic amines are important neurotransmitters. Norepinephrine is released by sympathetic nerves to control the heart and vascular smooth muscle.

Neuropeptides are small proteins released as neurotransmitters.

Synaptic release involves many proteins and is controlled by Ca_V channels, which are opened when an action potential invades the presynaptic terminal.

Axons have a microtubule-based transport system to move materials from the cell body to the presynaptic terminal (orthograde transport) and in the other direction (retrograde transport).

Postsynaptic potentials (PSPs) are excitatory (EPSPs) if they make the postsynaptic cell more likely to initiate an action potential and inhibitory (IPSPs) if they make it less likely.

Neuromuscular transmission is a well-studied example of synaptic transmission.

Hypocalcemia reduces the number of vessels released when an action potential invades the presynaptic terminal.

At the neuromuscular junction, the number of open channels is proportional to the concentration of acetylcholine squared times the effective number of acetylcholine receptor channels.

Several clinically important drugs act at the neuromuscular junction.

The number of vesicles released per action potential depends on the rate and pattern of arrival of the action potentials.

The autonomic nervous system has two synapses outside the central nervous system. The first is cholinergic; the second is either adrenergic or cholinergic.

In general, CNS synapses are similar to the neuromuscular junction, but they differ in many important ways.

In the CNS, several transmitters act through G-protein–coupled receptors to modulate the activity of the brain.

In order to fire repetitively, nerve cells use Ca-activated K channels to hyperpolarize the cell and allow the Na_V channels to recover from their inactivation.

Learning and memory involve changes in synaptic efficacy.

STUDY QUESTIONS

6–1. *Describe 10 steps in chemical synaptic transmission.*

6–2. *What is the effect on transmitter release if the extracellular calcium concentration is reduced? What proteins in the synapse are involved in the role of calcium?*

6–3. *Describe the generation of the endplate potential. What are the effects of esterase inhibitors and blockers of acetylcholine receptors?*

6–4. *Describe the flow of ions that produce EPSPs and IPSPs.*

6–5. *A medical student suffers a nerve crush injury of her ulnar nerve at the elbow. Calculate the approximate time required before she can expect to regain feeling in her fingertips.*

SUGGESTED READINGS

Draganski B, Gaser C, Busch V, et al. Neuroplasticity: Changes in grey matter induced by training. *Nature* 2004;427:311–312.

Freund TF, Katona I, Piomelli D. Role of endogenous cannabinoids in synaptic signaling. *Physiol Rev* 2003;83:1017–1066.

Hirokawa N, Takemura R. Molecular motors in neuronal development, intracellular transport and diseases. *Curr Opin Neurobiol* 2004;14:564–573.

Lalli G, Bohnert S, Deinhardt K, et al. The journey of tetanus and botulinum neurotoxins in neurons. *Trends Microbiol* 2003;11:431–437.

Schiavo G, Matteoli M, Montecucco C. Neurotoxins affecting neuroexocytosis. *Physiol Rev* 2000;80: 717–766.

Südhof TC. The synaptic vesicle cycle. *Annu Rev Neurosci* 2004;27:509–547.

WHO. *Neuroscience of Psychoactive Substance Use and Dependence.* Geneva: World Health Organization, 2004.

Muscle

7

OBJECTIVES

- *Discuss the proteins that compose the contractile apparatus of muscle.*
- *Describe the specializations of membranes and filaments for excitation-contraction coupling.*
- *Relate the structure of sarcomeres to their function.*
- *Discuss the structural and functional features that distinguish smooth muscle from striated muscle.*
- *Describe excitation-contraction coupling in smooth muscle.*
- *Describe single-unit and multiunit smooth muscles and their respective locations and functions.*
- *Discuss excitation-contraction coupling in cardiac muscle and explain how it differs from e–c coupling in skeletal and smooth muscle.*
- *Discuss the mechanics of cardiac contraction with reference to length-tension and pressure-volume diagrams.*
- *Describe the role of autonomic innervation and the effects of acetylcholine and norepinephrine on cardiac contractility.*

The outward expression of CNS activity is muscular contraction and glandular secretion. The nervous system controls three different types of muscles: **skeletal, cardiac**, and **smooth**. All three shorten and generate force via an ATP-consuming interaction between actin and **myosin**. However, they are cells of varying shape with different patterns of innervation and different mechanisms of excitation-contraction coupling.

Skeletal and cardiac muscles are also called **striated**, or striped, because of the characteristic pattern seen in the light microscope—a pattern that is lacking in smooth, **nonstriated** muscle (Fig. 7-1). Skeletal muscles are long multinucleate cylinders 20 to 100 μm in diameter and several centimeters in length. Cardiac muscle cells are mononucleate, about 20 μm by 50 μm long, and they may be branched. They are connected end to end by cell adhesion molecules and also cell-to-cell channels, so the heart functions as an electrical syncitium.

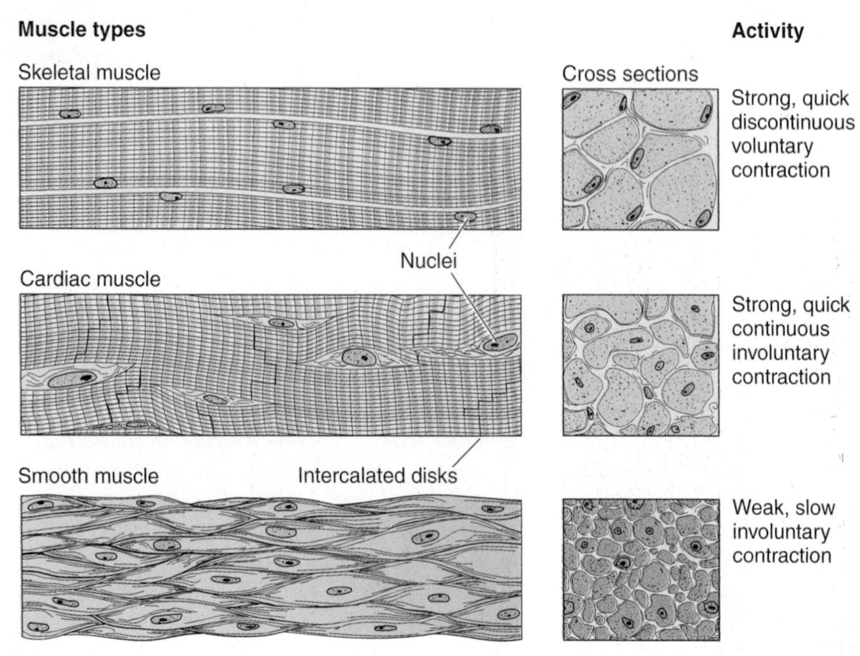

Figure 7-1. The three types of muscles. Skeletal muscle is composed of large, elongated multinucleated fibers. Cardiac muscle is composed of irregular branched cells bound together longitudinally by intercalated disks. Smooth muscle is an agglomerate of fusiform cells. The density of the packing between the cells depends on the amount of extracellular connective tissue present. (From Junqueira LC, Carneiro J. *Basic Histology: Text & Atlas*, 11th ed. New York: McGraw-Hill, 2005, with permission.)

Smooth muscle cells are mononucleate, about 5 μm in diameter by 20 μm long. In the gut, blood vessels and uterine walls, the smooth muscle cells are coupled by cell-cell junctions, and many cells work together as a unit. These are called **unitary** or **visceral** smooth muscles to distinguish them from **multiunit** smooth muscles, found in the iris and ciliary body in the eye, and piloerector muscles at the base of hair follicles. These two types, unitary and multiunit, represent the ends of a continuous spectrum. The individual cells of multiunit smooth muscles receive innervation allowing for finer control. In visceral smooth muscle, one nerve may control many muscle cells via cell-cell channels.

FORCE GENERATION AND SHORTENING

The **length** of a muscle cell, the **velocity** with which it is shortening, and the force or **tension** within the muscle are three independent parameters that characterize the mechanical state of the cell. There are relationships between them that ultimately derive from the interaction between actin and myosin. It is convenient to speak of two idealized activities of muscle, **isometric** and **isotonic** contraction.

Isometric (from the Greek meaning "same length") contractions are increases in force generated by a muscle that is not allowed to change its length and has zero velocity of contraction. This happens if you attempt to lift something that is bolted to the floor. The amount of force generated by the actin-myosin interaction depends on the length of the muscle. Isotonic ("same tension") contractions are shortenings of the muscle against a constant load, as when you lift a spoon or a book. Isotonic contractions occur at a constant velocity; there is a relationship between the force produced and the velocity of contraction.

Most skeletal muscles are arranged across a joint; other **antagonist** muscles move the joint in the opposite direction. When you contract your biceps and your hand approaches your shoulder, your triceps are being passively stretched. There is no active elongation of muscles. The passive stretch produces tension within the triceps because elements of the triceps muscle resist its tendency to break. The total tension within a muscle is the sum of **passive tension** due to forces outside the muscle and **active tension** produced by actin-myosin interactions inside the muscle. In the heart, passive stretching and passive tension occur as the chambers fill with blood between contractions.

The functional unit of skeletal and cardiac muscle contraction is the **sarcomere** (Fig. 7-2), which is repeated many times in series and underlies the striped pattern. The thick myosin filaments and the less dense areas between them that contain the thin actin filaments and the Z lines form the most obvious stripes. The sarcomere is about 2 μm from Z line to Z line; during contraction, the **thick** and **thin filaments** slide over one another and the **Z lines** come closer together. The region of the thick filaments is often called the **A band** because of the ordered quality or anisotropy of myosin. The **I band**, ("I" for *isotropic*) extends from one A band to the next, with half an I band in each sarcomere. The I bands are ordered, but less so than the A bands. When the muscle shortens, the A bands remain the same length and the I bands become shorter.

The sarcomere contracts because there are **cross bridges** between the thick and thin filaments. The myosin molecule may be functionally divided into three regions, the **head**, the **lever arm**, and the **tail**. The cross bridges are

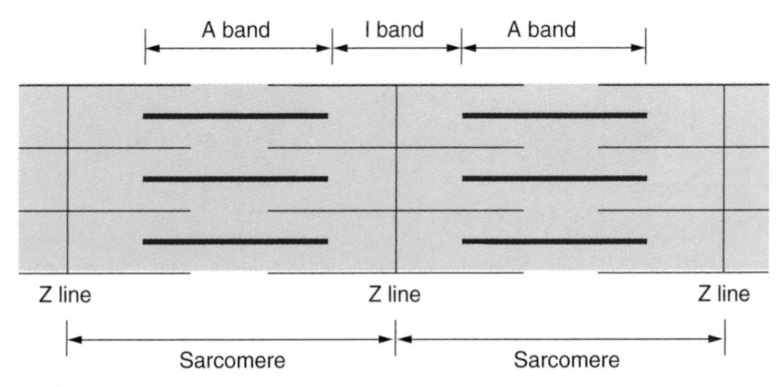

Figure 7-2. A schematic drawing of two sarcomeres from a striated muscle.

globular head domains of the myosin molecules connected by an 8.5-nm lever arm to the core of the thick filaments, which consists of myosin tails. The head can undergo a cyclic **mechanochemical reaction** with the actin filament (Fig. 7-3). Starting from the attached state, the binding of ATP to the myosin head will release it from the actin. Myosin has a low affinity for actin when ATP or ADP and inorganic phosphate, Pi, are bound to myosin. Splitting the ATP into ADP and Pi, both still bound to the myosin, causes the myosin to go into the cocked state. Upon Pi release, the affinity of myosin for actin increases. Binding again to the actin is associated with a conformational change in the myosin and force generation in a 5-nm power stroke. The ADP is then released and the myosin returns to the attached

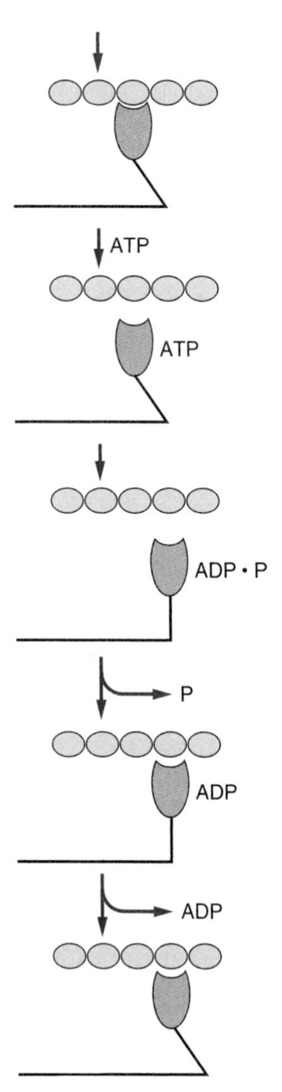

Figure 7-3. The mechanochemical reaction cycle.

state. In the absence of ATP, the myosin will remain attached to the actin, a condition known as **rigor**. The stiffness seen a few hours after death, **rigor mortis**, is due to this attachment.

If the muscle is allowed to shorten, the Z lines will draw closer together and each attachment will take place a little closer to the Z line. The movement is smooth because the myosin heads are not synchronized. The power stroke takes only about 5 percent of the cycle time; most of the time is passed in the unattached states. This allows many myosins to work together to move the actin. If the sarcomere is kept at a constant length, the myosin head will exert a force of about 5 pN and stress the bonds holding the actin filament together. During the released portion of the cycle, the strain will be relieved and each successive attachment will, on average, occur at the same location. If the load is very small, the muscle will contract at its maximal rate, which is determined by the time it takes to go through the ATP hydrolysis cycle (Fig. 7-3). If the load is greater, some of the myosin heads will be holding the load and the rate of contraction will be less than maximal. When the load is so great that all of the myosin heads are required to hold it, the velocity of contraction will be zero and the muscle will be isometric. Thus the velocity of contraction is inversely proportional to the load. This corresponds to the experience that you can move a light object more rapidly than a heavy one.

The isometric tension that a muscle can produce is related to the sarcomere's length (Fig. 7-4), because this determines how many myosin head groups will find a suitable actin to bind. If the sarcomere is stretched to about twice its usual length in the body, there will be no overlap of thick and thin filaments and no tension can be generated. If the muscle attempts to become much shorter than its usual length, the amount of tension it can produce will be less because the overlapping actin filaments from the two ends interfere with myosin binding. If the muscle becomes short enough, the Z lines will meet the thick filaments, which will prevent further shortening. In the healthy body, the joints limit

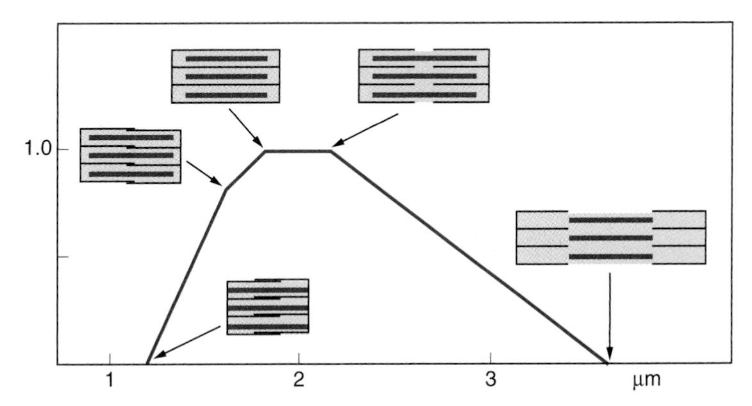

Figure 7-4. The isometric length-tension diagram.

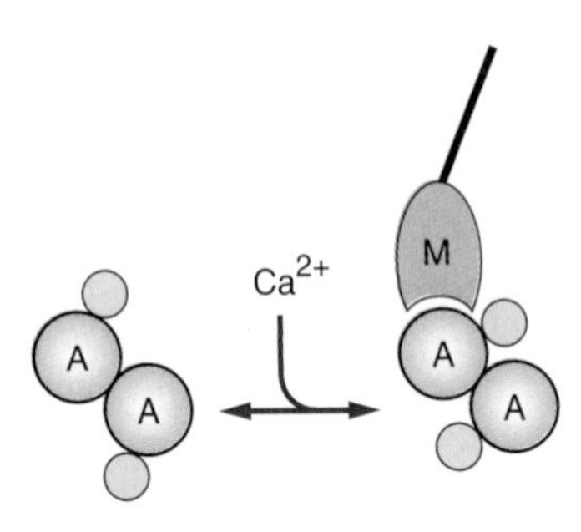

Figure 7-5. Troponin-tropomyosin controls the availability of myosin binding sites on actin in striated muscle.

the extension and contraction of the muscles, so they remain near their optimal sarcomeric length.

Along with actin and myosin, other proteins have important functions in the sarcomere. **Troponin** and **tropomyosin** are in the thin filaments and regulate contraction (Fig. 7-5). **Titin**, the largest known protein (about 3700 kDa), extends about 1 μm from the Z line to the midline of the sarcomere. It lies between thin filaments and is imbedded in the myosin of the thick filaments. Titin provides some of the elasticity of muscles and serves to center the thick filaments between the Z lines and also between the thin filaments. **Nebulin** is another giant protein (500 to 900 kDa) that is associated with the thin filaments. There are other proteins that make the Z lines (e.g., α-**actinin**), cap the thin filaments, and make the mid-sarcomeric structure known as the M line.

The contraction of smooth muscles depends on the same cyclic mechanochemical interaction between actin and myosin. However, there are no sarcomeres and the actin and myosin filaments are not all aligned with the long axis of the cells. There are **dense bodies** near the surface of the cell that contain α-actinin and serve the purpose of the Z lines in connecting to actin filaments (Fig. 7-6).

5 Muscle cells initiate a contraction when the cytoplasmic Ca concentration increases; they relax when the Ca is removed. The control of this Ca concentration is discussed further below. In striated muscle, Ca controls the contraction by an effect of the thin filaments. The filamentous actin polymer resembles two strings of pearls twisted gently around each other. Tropomyosin is a long molecule that lies in the grooves between the strings for a distance of seven actin monomers. In the absence of Ca, tropomyosin prevents the actin-myosin

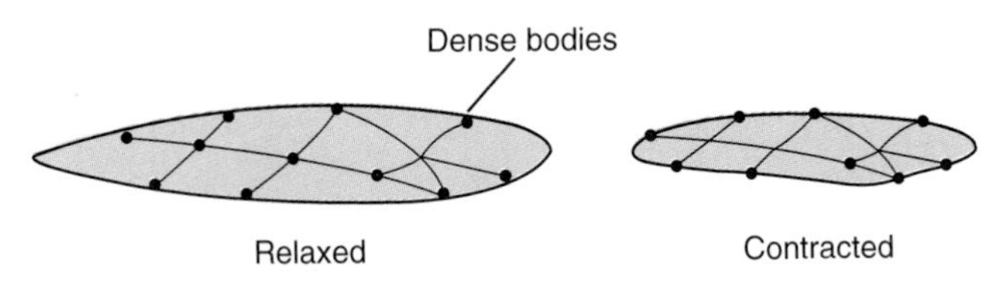

Figure 7-6. A smooth muscle cell contracting.

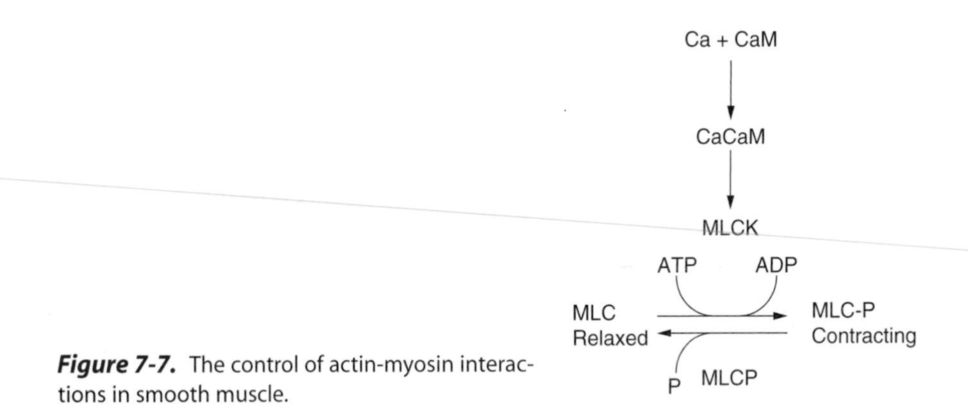

Figure 7-7. The control of actin-myosin interactions in smooth muscle.

interaction and the muscle is relaxed. Troponin is a more globular molecule that binds to tropomyosin and has Ca binding sites. When Ca binds to troponin, it moves the tropomyosin, the actin binding sites for myosin are exposed, and the mechanochemical cycle is permitted. Skeletal and cardiac muscles both use this method, although cardiac troponin is a different gene product.

In contrast, Ca regulates smooth muscle contraction by an effect on the heavy filaments. Smooth muscle lacks troponin. Ca initiates contraction by first binding with **calmodulin** (**CaM**). The **Ca-CaM** complex activates **myosin light-chain kinase** (**MLCK**), which phosphorylates the **regulatory** myosin light chain, thus permitting the actin-myosin reaction (Fig. 7-7). There are also myosin light-chain **phosphatases**, which permit relaxation.

Almost all cells, including striated muscle, contain calmodulin. The Ca-CaM complex is used in many other circumstances. Both striated and smooth muscle myosins have associated **myosin light chains**, or small accessory proteins. Two different light chains are wrapped around each lever arm. One is called **essential**, because it was required in reconstitution experiments, and the other is called **regulatory**.

Smooth muscle also has the ability to enter a "**latch**" state, which allows the maintenance of tone with minimal ATP consumption. The basis of the latch hypothesis was the observation that myosin phosphorylation in smooth muscle was not correlated with force but with shortening velocity. In the latch state, shortening is not occurring; the latch state is analogous to the rigor state of skeletal muscle. It is thought that the latch state occurs by dephosphorylation of the myosin light chain while the myosin is attached to the actin and that, under these circumstances, the tension can be maintained. A slow return to the unattached state is also postulated.

CONTROL OF INTRACELLULAR CALCIUM

Calcium controls the contractile machinery; channels, pumps, and transporters control the calcium. The details of this control are different for the different types of muscle, although there is some overlap between types.

In both skeletal and cardiac muscle, an action potential triggers contraction. In skeletal muscle, the action potential is a brief (about 2-ms) event that propagates rapidly over the surface of the muscle and down a transverse network of invaginations, the system of t tubules (Fig. 7-8). There are molecules in the surface membrane of the t tubules with a structure similar to that of Ca_V channels. These are called **dihydropyridine receptors** (**DHPR**), after a class of drugs that inhibit them. DHPRs have the 4-by-6 TM topology of Ca_V channels and undergo a conformational change when depolarized; in skeletal muscle t tubules, however, they do not allow much Ca into the cell. Instead, the DHPRs induce a conformational change in **calcium release channels** (**CRCs**) in the membranes of the adjacent sarcoplasmic reticulum. The CRCs are large proteins (565 kDa) that form tetrameric Ca channels. These CRCs are also called **ryanodine receptors** (**RyR**) because their activity is altered by ryanodine, the principal alkaloid of the botanical insecticide ryania. After it is released, the SERCA pump pumps the Ca back into the sarcoplasmic reticulum and the muscle relaxes. In the lumen of the sarcoplasmic reticulum, much of the Ca is bound to a Ca-binding protein called **calsequestrin**. This binding is reversible; the Ca is released by the calsequestrin when the RyRs allow Ca to flow into the cytoplasm.

Cardiac muscle differs from skeletal by having prolonged action potentials (about 100 to 200 ms), smaller cell diameters, and a less developed t-tubule system. Cardiac DHPRs permit Ca entry; they are also called L-type Ca_V channels because they remain open a long time. This Ca entry helps keep the membrane depolarized during the plateau phase of the action potential. The next step is a **Ca-induced Ca release** (**CICR**) through the cardiac RyRs. which supplies most

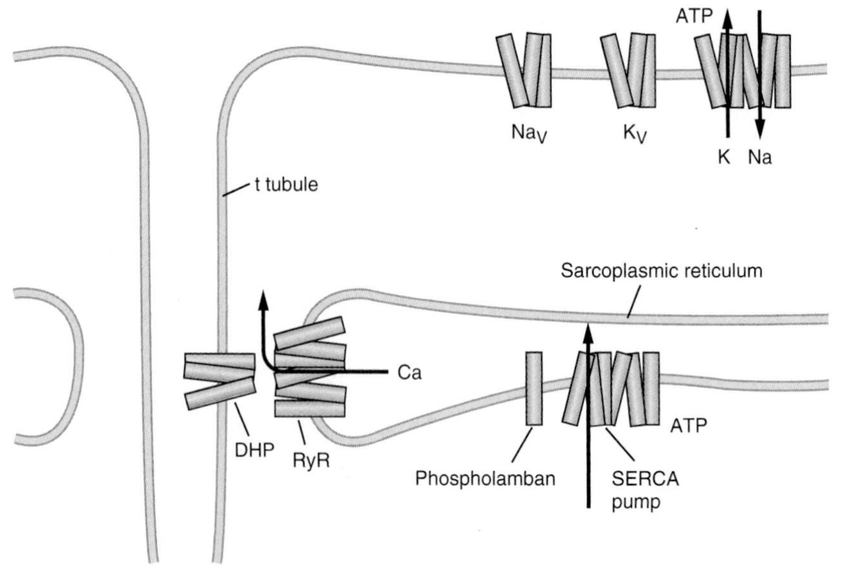

Figure 7-8. The control of intracellular calcium in striated muscle.

of the Ca that binds the troponin. Unlike skeletal muscle, the heart will fail to contract in the absence of extracellular Ca. When the membrane repolarizes after the action potential, SERCA pumps pump most of the Ca back into the sarcoplasmic reticulum, but Ca pumps and Na/Ca exchangers pump some out of the cell through the surface membrane. Stimulation of β-adrenergic receptors on cardiac muscle cells activates a cAMP, PKA sequence that enhances contractility by increasing the influx of Ca through L-type Ca_V channels. This system also enhances the relaxation rate by phosphorylating **phospholamban**, a regulator of the SERCA pump.

The tasks that are carried out by smooth muscle are more diverse than those of skeletal and cardiac muscle, and so are the ways in which intracellular Ca is controlled. The most prominent pathway for activation of contraction in vascular smooth muscle is via norepinephrine binding the **α-adrenergic receptor** (a GPCR) and via $G\alpha_q$ subunits stimulating phospholipase C (**PLCβ**) to make IP_3. The IP_3 binds to IP_3 receptors on the sarcoplasmic reticulum. IP_3 receptors are similar to RyRs in that they are large tetramers and open to allow Ca to flow down its electrochemical gradient into the cytoplasm. A SERCA pump is present to recover the Ca and allow the muscle to relax. There are IP_3 receptors and SERCA pumps on the endoplasmic reticulum in most cells. There are also RyRs in the sarcoplasmic reticulum of some smooth muscle cells; they are thought to participate in CICR.

Many smooth muscle cells have L-type Ca channels in their surface membrane and respond to depolarization by allowing Ca entry, which may induce Ca release from the sarcoplasmic reticulum. Some smooth muscle cells have action potentials, some have slow oscillatory fluctuations of membrane potential with periods of several seconds, and some do not change their membrane potential under normal circumstances (but all will depolarize if the external K concentration increases).

There are Ca-permeable nonselective cation channels in some vascular smooth muscle cells that are opened when the Ca in the sarcoplasm is depleted by previous activity. These are called **store-operated channels (SOCs)**, "store" referring to the sarcoplasmic reticulum's storage of Ca. The channels are defined functionally; their molecular identity and the link between store depletion and channel opening are not known. In portal vein smooth muscle cells, it has recently been shown that NE can also control SOCs, possibly by a store-independent mechanism.

There are many agents that promote relaxation of vascular smooth muscle by changing the cAMP or cGMP levels. Nitric oxide (NO) stimulates a **cytoplasmic guanylyl cyclase,** which increases cGMP-stimulating phosphokinase G and causes relaxation by activating myosin light-chain phosphatase. Activation of β_2 adrenergic receptors or H2 histamine receptors causes relaxation by elevating cAMP, which leads to inhibition of MLCK via phosphokinase A.

MECHANICAL OUTPUT

Muscles are designed to do mechanical work. Skeletal muscles move the body in the gravitational field and move external objects. Cardiac muscle pumps blood and elevates blood pressure. Vascular smooth muscle regulates the diameter of blood vessels, thereby controlling the flow through them. Each type of muscle has different strategies to control its mechanical output.

Skeletal muscles are driven by motoneurons; there is a one-to-one correspondence between a motoneuron action potential and a twitch contraction of all of the skeletal muscle cells that receive its synaptic terminals. These motor units (the motoneuron and all of its muscle cells) come in various sizes, from three to several hundred muscle cells. The CNS controls a muscle by choosing which motor units to excite, choosing the smaller motor units first when less strong but more controlled forces are required.

The CNS does not drive the motor units with single impulses, which would cause twitch contractions, but rather with trains of impulses at 20 to 50/s. At these rates the individual twitch contractions are summed together into a **fused tetanic contraction** (Fig. 7-9). The summation occurs because the mechanical events are much slower than the electrical events and the muscle can be stimulated again before all of the Ca has been pumped back into the sarcoplasmic reticulum. If the rate is high enough, the troponin will remain saturated with Ca and the contraction will be continuous (Fig. 7-10).

The force that the muscle applies to the bone during tetanic stimulation is larger than that seen during a single twitch (Fig. 7-9), even though all of the troponin was activated in both situations. This is because there are **elastic elements** in series with the contractile elements (Fig. 7-11). The tendons at the ends of the muscle are a large part of this **series elasticity**, but any tendency to stretch in the thin filaments, the Z lines, and the attachment of the last Z lines to the tendon would also contribute. The cross bridges pull on the series elastic elements and the series elastic elements pull on the load. It takes about 30 ms to stretch the series elasticity so the full force of the cross bridges is not transmitted to the load during a single twitch.

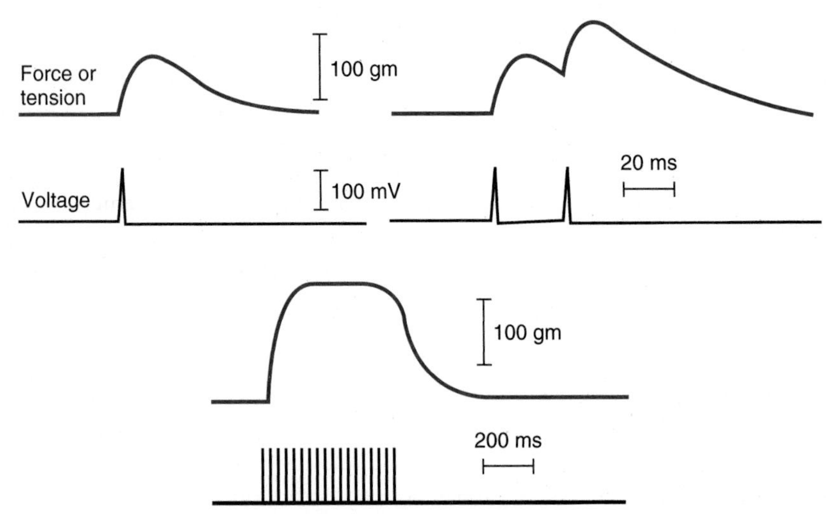

Figure 7-9. The temporal summation of mechanical responses in skeletal muscle.

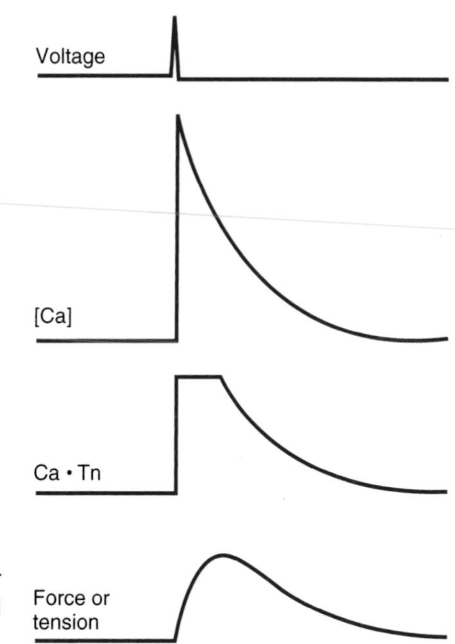

Figure 7-10. The time course of cellular events during a twitch contraction of skeletal muscle.

Figure 7-11 also shows elastic elements in parallel with the contractile elements. Extracellular fascia surrounding the muscle cells contribute most of the **parallel elasticity**, but titin molecules also act in parallel with the cross bridges. The parallel elasticity determines the behavior of the muscle in response to external forces when the muscle cell is not stimulated. Figure 7-12 shows a simple experiment to measure both the passive and active length-tension relationship. As the muscle is passively stretched, the tension resisting this stretch increases, just as it does when you stretch a string or elastic band. When the muscle is stimulated, it produces active tension that depends on the cross-bridge overlap (Fig. 7-4). The total tension in the muscle is the sum of the active and passive tension. Excessive total tension can tear the muscle or the junction between tendon and bone.

Muscles in the body vary with regard to the amount and **compliance** (the reciprocal of stiffness) of their connective-tissue parallel elastic elements. They also vary in the speed of contraction and ability to resist fatigue. There are different types of skeletal muscle fibers with different genes expressing different versions

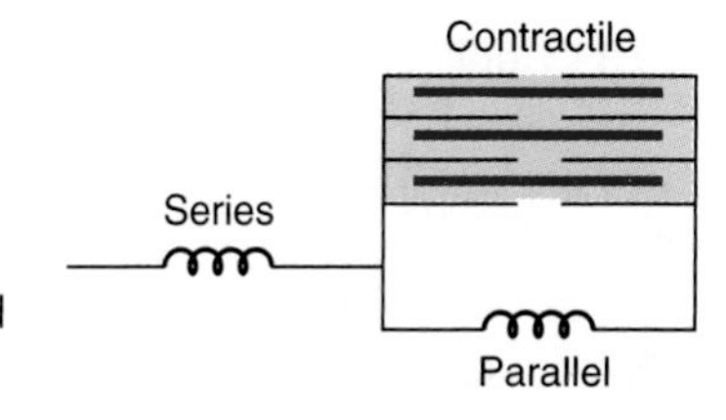

Figure 7-11. A schematic drawing of series and parallel elastic elements of muscles.

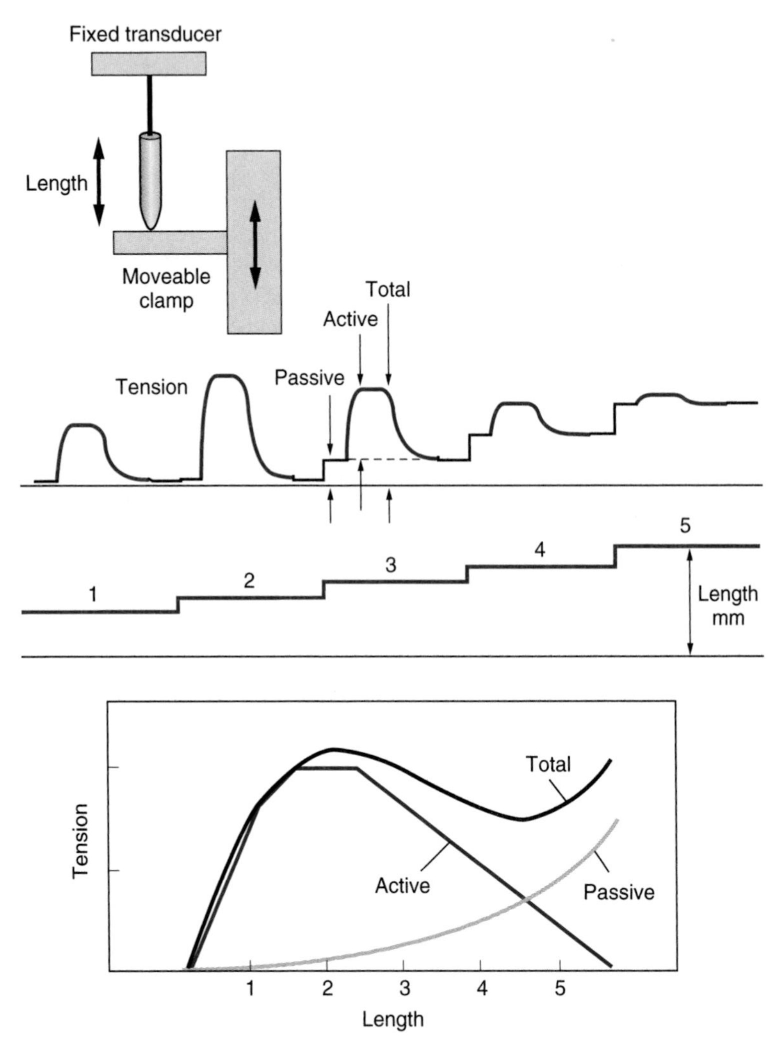

Figure 7-12. The measurement of passive, active, and total tension in muscles as a function of muscle length.

of myosin (both heavy and light chains), troponin, and the SERCA pump. The faster muscles—e.g., the gastrocnemius—are also more fatigable and are used for rapid motion. The soleus is a slower muscle; it is less fatigable and is used to maintain posture. The fast muscles have larger-diameter fibers and less myoglobin in their cytoplasm, thus they are "**white**" as opposed to the slow, smaller fibers of the "**red**" muscles. The slow muscles have more mitochondria and rely on oxidative metabolism. The fast muscles have greater glycogen stores and often depend on glycolytic metabolism.

All skeletal muscles have several layers of backup to maintain the ATP levels required for contraction. The fastest recovery comes from transferring a high-energy phosphate from **creatine phosphate** to ADP by **creatine**

phosphokinase (**CPK**), an important muscle enzyme. Creatine phosphate cannot be directly hydrolyzed by myosin, but it can serve as a reservoir of readily available high-energy phosphate. **Myokinase**, another important muscle enzyme, catalyzes the reaction 2 ADP –> AMP + ATP, which also helps to maintain ATP levels.

Glycolysis, the breakdown of stored glycogen to pyruvate, is the next level of backup. Glycolysis is stimulated by a rise in cytoplasmic Ca, which activates **phosphorylase kinase;** this, in turn, activates **phosphorylase**, which produces glucose-1-phosphate, the first step of glycolysis. In the absence of sufficient aerobic metabolism by the mitochondria, pyruvate will be converted to **lactate,** which will accumulate and leave the muscle via proton-linked monocarboxylate transporters. Once outside, it may produce pain by stimulating acid-sensing ion channels in nerve endings. In slow muscles, **myoglobin** stores oxygen in the cytoplasm, and the larger number of mitochondria ensures that oxidative phosphorylation will be available for more endurance without lactate production.

Along with their mechanical work, active skeletal muscles generate heat. The heat may be useful in cold environments; shivering is one of the ways the body uses to maintain a constant internal temperature. There are some people, however, who are in danger of experiencing **malignant hyperthermia** if they are subjected to general anesthetics such as halothane. The symptoms include muscle rigidity and high fever. In some families a genetic linkage with the RyR has been found. In these patients, halothane triggers prolonged opening of the Ca release channels. The primary treatment is with **dantrolene**, which is thought to inhibit the RyRs directly or indirectly.

9 Cardiac muscle cells contract rhythmically, and every cell contracts with each heartbeat. The heart may be faced with more or less filling and will vary its volume output to match its input. The function of the heart is often described with **pressure-volume** curves (Fig. 7-13) that are directly related to the length-tension diagram (Fig. 7-12). As the ventricle fills with blood (1) its volume increases and the length of the muscle cells increases, producing a passive tension in the cells and the diastolic (relaxed) pressure in the ventricle. When the muscle contracts, the force and pressure increase, at first (2) isometrically (or **isovolumetrically**), because both the valves leading into and out of the ventricle are closed. When the pressure in the ventricle exceeds the pressure in the aorta, the

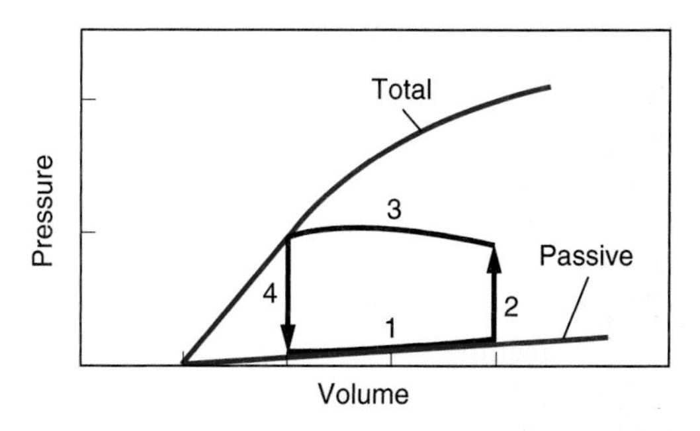

Figure 7-13. The pressure-volume loop of the cardiac cycle.

aortic valve opens and blood leaves the ventricle (3). The muscle fibers are now shortening under an almost constant load, that is, almost isotonically. They will continue to shorten until they reach a length where they have just enough active tension to match the pressure load. At the end of the cardiac action potential, the muscle relaxes (4), again isovolumetrically, because both valves are closed. When the pressure in the ventricle falls below the pressure in the atrium, the atrioventricular valve opens and the ventricle fills again.

The body has three ways to vary cardiac function. There can be more filling, increasing the "**preload**" and stretching the resting fibers further. The heart will respond by ejecting more blood, reaching the same point determined by the active length-tension relationship. Second, aortic pressure can go up, increasing the "**afterload,**" thus making the fibers work against a higher load and reducing the amount of blood that is ejected. Third, the sympathetic nervous system can change the active length-tension relationship. Norepinephrine can increase "**contractility**," or the maximum tension that can be achieved at any length. In this case, if the filling maintains the starting volume, more blood will be ejected on each stroke. All of these effects are reversed if the inputs are decreased.

Norepinephrine increases contractility via a PKA phosphorylation of the L-type Ca_V channel, so more Ca enters, more is released from the sarcoplasmic reticulum, more binds troponin, and more cross bridges are made at any length. Unlike skeletal muscle, cardiac troponin is not saturated during every contraction. Also in cardiac muscle, unlike skeletal muscle, the force delivered to the load cannot be increased by tetanic stimulation. The cardiac action potential and the contraction have similar time courses. Unlike the skeletal muscle, the cardiac muscle is electrically refractory for the duration of the contraction.

As in skeletal muscle cells, there is diversity among cardiac muscle cells. Action potentials in ventricular cells are longer than in atrial cells. Inner (endocardial) ventricular cells have longer action potentials than outer (epicardial) cells. These differences are due to differing amounts of the various ionic channels. The duration of contractions follows the duration of the action potentials. The electrical properties of the heart's various cells are apparently tuned to control optimal cooperative functioning.

Heart muscle works continuously, with a duty cycle of about 30 percent; the duration of systole is about one-half the duration of diastole. There are many mitochondria as well as a relatively high concentration of myoglobin to support aerobic metabolism. Heart tissue requires a continuous supply of oxygen to provide the ATP for contraction and the maintenance of ionic gradients. A local loss of oxygen due to a blocked blood vessel (**ischemia**) will lead to loss of contraction in seconds and permanent local tissue damage (**infarct**) in minutes. If the region of loss is large, the result may be fatal. In normal circumstances, about three-quarters of cardiac ATP is produced by the breakdown of circulating lipids, although other sources can be used if lipids are not available. Lipid metabolism produces the most ATP per gram, more than twice as much as glucose.

Smooth muscles are much more diverse than skeletal and cardiac muscles. Some smooth muscles are contracted almost all of the time and some very infrequently. The lower esophageal sphincter at the junction between the esophagus and the stomach is closed most of the time to prevent acid reflux. The smooth muscles in the walls of the urinary bladder are relaxed except during urination. The smooth muscles of the intestinal walls are either relaxed or undergoing rhythmic contractions in waves of peristalsis that propagate down the intestine, propelling its contents. The muscles determining pupil size in the eye and the diameter of arterioles everywhere have a steady resting contraction and the capability to respond to hormones or transmitters with either contraction or relaxation. The muscles of the uterine myometrium are relaxed most of the time; weak, irregular contractions are seen only in the last month of pregnancy, and strong contractions only at parturition.

At the initiation of parturition, there is a dramatic increase in the number of gap junctions between myometrial cells. During the second half of pregnancy, the number of oxytocin receptors increases 200-fold, with half of that increase in the last 4 weeks. When labor is initiated, oxytocin is released into the blood from neurosecretory cells, with their cell bodies in the hypothalamus and their terminals in the posterior pituitary. The myometrial oxytocin receptors are GPCRs that activate contraction. The primary stimulus for the release of oxytocin appears to be cervical distention, which means that parturition is an example of a positive feedback loop.

Not only are smooth muscles diverse, but there are other cells that exhibit actin-myosin–based contractility. Pericytes are solitary smooth muscle–like cells that surround the walls of some microscopic blood vessels. Fibrocytes are cells of loose connective tissue that have been shown to contract in tissue culture and play a role in wound healing. Cell migration involves an actin-myosin interaction, and cell division is accomplished by a transient formation of the contractile ring, made of actin and myosin, which constricts to separate the two daughter cells.

KEY CONCEPTS

Functionally and histologically there are three categories of muscle cells: skeletal, cardiac, and smooth.

When activated, muscles shorten and/or produce force or tension. Muscle contraction is often idealized as isometric or isotonic.

The functional unit of skeletal and cardiac muscle contraction is the sarcomere. In all three types of muscle, a cyclic interaction between myosin head groups and actin filaments produces the contraction.

Isometric tension is related to sarcomeric length because of the varying overlap of the thick, myosin-containing and thin, actin-containing filaments.

The cytoplasmic calcium concentration controls the interaction between actin and myosin.

In skeletal and cardiac muscle, DHP receptors control the release of calcium from the sarcoplasmic reticulum by different mechanisms.

A motor unit is a motoneuron and all of the skeletal muscle cells to which it connects.

There are several backup systems to maintain ATP levels during muscle activity.

The pressure-volume relationship seen in the functioning heart is similar to the length-tension relationship for individual cells.

Smooth muscle cells are diverse.

STUDY QUESTIONS

7–1. Define isometric contraction and isotonic contraction.

7–2. Compare the time course of an isometric mechanical "twitch" contraction with the muscle action potential.

7–3. What is the mechanical response to repetitive stimulation of a nerve leading to a muscle?

7–4. What is the mechanism of "tetanic fusion"?

7–5. How does the muscle tension seen during an isometric tetanus vary with the length of the muscle? What is the molecular basis for this relationship?

7–6. How does the resting tension in the muscle vary with muscle length?

7–7. How does the velocity of an isotonic contraction vary with the load?

7–8. Compare the properties of skeletal and cardiac muscle contraction.

7–9. Compare activation of skeletal muscle and smooth muscle.

SUGGESTED READINGS

Baker JE, Brosseau C, Fagnant P, Warshaw DM. The unique properties of tonic smooth muscle emerge from intrinsic as well as intermolecular behaviors of myosin molecules. *J Biol Chem* 2003;278:28533–28539.

McElhinny AS, Kazmierski ST, Labeit S, Gregorio CC. Nebulin: The nebulous, multifunctional giant of striated muscle. *Trends Cardiovasc Med* 2003;13:195–201.

Parekh AB, Putney JW Jr. Store-operated calcium channels. *Physiol Rev* 2005;85:757–810.

Rembold CM, Wardle RL, Wingard CJ, et al. Cooperative attachment of cross bridges predicts regulation of smooth muscle force by myosin phosphorylation. *Am J Physiol Cell Physiol* 2004;287:C590–C602.

Toyoshima C, Inesi G. Structural basis of ion pumping by Ca^{2+}-ATPase of the sarcoplasmic reticulum. *Annu Rev Biochem* 2004;73:269–292.

Wehrens XH, Lehnart SE, Marks AR. Intracellular calcium release and cardiac disease. *Annu Rev Physiol* 2005;67:69–98.

Answers to Study Questions

CHAPTER 1

1–1. See Fig. 1-2.
1–2. See Fig. 1-3.

CHAPTER 2

2–1. In free diffusion, the flux is proportional to the concentration gradient at all concentrations. Facilitated diffusion is characterized by a maximum flux rate and a concentration where the flux is half-maximal. Primary active transport moves materials against the concentration gradient at the expense of ATP hydrolysis. Secondary active transport moves some materials up their concentration gradient at the expense of moving other materials down theirs.

2–2. By providing an environment within the channel that mimics the water environment of the ion in bulk solution—i.e., that has similar charges at similar positions and distances from the ion.

2–3. If she drinks 4 L of distilled water, the volume of all of her body compartments will increase by 10 percent (4/40) and their osmolarity will decrease by 10 percent. If she drinks 1 L of isotonic solution, her extracellular volume will increase by 1 L, her intracellular volume will not change, and the osmolarities will not change.

2–4. Gap junctions allow the movement of small molecules between adjacent cells. In some conditions—e.g., damage to a cell—it is preferable to prevent this movement.

2–5. To assure that the appropriate cells will connect to each other and the inappropriate ones will not.

CHAPTER 3

3–1. Electrical current is the net flux of charge. The direction of electrical current is the same as the direction of flux of positive ions (or opposite to the flux of negative charge). One ampere of current is the flux of 1 C/s or 1/96,484 mol of monovalent cations per second.

3–2. The resting potential is a separation of charge across the cell membrane. It is produced primarily by the efflux of K ions down their concentration gradient through K channels, leaving excess negative charge inside the cell. The resting potential is slightly less negative than would be expected if K ions were in electrochemical equilibrium, because there are also a few open Na channels.

3–3. (a) A small depolarization (about 2 mV).
(b) A small hyperpolarization.
(c) A large depolarization (about 30 mV).
(d) No effect.
(e) A small hyperpolarization.

3–4. An immediate very small (about 1 mV) depolarization followed by a very slow (hours) depolarization.

3–5. –94 mV, +62 mV, −15 mV.

3–6. See Fig. 5-1.

CHAPTER 4

4–1. The appropriate stimulus for a given receptor—e.g., light for the eye.

4–2. The region in stimulus space that evokes a response in a specified cell.

4–3. The conversion of mechanical energy or photons or the presence of chemicals into a change in membrane potential.

4–4. No. Photoreceptors respond to light with a hyperpolarization (by the reduction of an inward current).

4–5. Sensory adaptation refers to a sensory response that decreases in time when presented with a steady maintained stimulus. It occurs in all sensory systems except pain and arises at many levels between the stimulus and the response.

4–6. A communication path where the receiver knows the type of information by knowing on which line it arrives.

CHAPTER 5

5–1. Inward Na current and outward K current.

5–2. a, increase; b–e, decrease

5–3. During the absolute refractory period, the nerve cannot be excited. During the relatively refractory period, the nerve can be excited, but it requires a larger stimulus that normal. Both reflect the recovery of Na_V channels from inactivation experienced during a previous action potential. During the absolute refractory period, so few channels have recovered that even if all were opened, the resulting inward Na current would not exceed the outward K current.

5–4. Cable properties are passive properties that govern the rate of change of membrane potential and its spread along the length of a cell. The membrane capacitance, the resting membrane resistance, and the longitudinal axoplasmic resistance are the component parts of the cable properties.

5–5. See Fig. 5-1.

CHAPTER 6

6–1. An action potential enters the presynaptic terminal. The nerve terminal is depolarized. Depolarizing opens Ca_V channels. Ca enters the cell, moving down its electrochemical gradient. Ca acts on synaptotagmin, causing synaptic vesicles to fuse with the presynaptic membrane. The vesicles release neurotransmitter into the synaptic cleft. The transmitter in the cleft (a) diffuses away out of the cleft, (b) is hydrolyzed or taken up by adjacent cells, or (c) interacts with receptors on the postsynaptic membrane. The activated receptors are permeable to particular ions. A flux of ions into or out of the postsynaptic cell changes the membrane potential in the synaptic region. If the postsynaptic cell is depolarized to threshold, an action potential is elicited and propagates in both directions to the ends of the postsynaptic cell. The transmitter is recycled into the presynaptic terminal, Ca is pumped out of the presynaptic terminal, and vesicles are recycled and refilled.

6–2. Lowering extracellular Ca reduces transmitter release. During the synaptic release cycle, Ca_V channels, the cell membrane Ca pump, and synaptotagmin interact with Ca. Ca also has interactions with mitochondria and cytoplasmic proteins.

6–3. An endplate potential is produced when about 200 vesicles of ACh are released from the nerve terminal. Some of the ACh binds to the postsynaptic AChRs, causing them to open and allow Na to enter the cell, thus producing a depolarization. Some of the ACh is hydrolyzed by ACh esterase in the cleft. Inhibition of the Ach esterase will increase the magnitude and prolong the endplate potential. Blocking of AChRs will reduce the magnitude of the endplate potential.

6–4. Opening channels with a reversal potential that is more depolarized than the threshold for the generation of action potentials produces EPSPs. The IPSPs come from channels with a reversal potential more negative than this threshold. When channels open, the membrane potential tends toward the reversal potential of the channel.

6–5. The average distance from the elbow to the fingertips is 1 cubit, about 18 in., or about 45 cm. At 1 mm/day, regeneration would take 450 days or about 15 months.

CHAPTER 7

7–1. In an isometric contraction, muscle length does not change but muscle tension is produced. In an isotonic contraction the force exerted by the muscle is constant and the muscle shortens.

7–2. The duration of a muscle action potential is about 1 ms. The duration of a muscle contraction is 10 to 100 ms or even slower, depending on the type of muscle.

7–3. Repetitive stimulation at 20 to 50/s will generally produce a fused tetanic contraction.

7–4. Fusion takes place because the next stimulus occurs before the intracellular Ca level has dropped below that required to saturate the troponin.

7–5. See Fig. 7-4. The number of cross bridges that can be made varies with the amount of overlap of the actin and myosin filaments.

7–6. See Fig. 7-12.

7–7. Inversely, higher velocity for lower loads, higher force at lower velocities. (Maximum force at zero velocity or isometrically.)

7–8. Skeletal muscle is generally either at rest or contracting tetanically, with all of its troponin bound by Ca. Cardiac muscle contracts in response to individual action potentials, and the amount of bound troponin may vary. Both have qualitatively similar length-tension and force-velocity relationships.

7–9. Skeletal muscle is activated by action potentials, which cause a conformational change in DHP receptors, which trigger the release of Ca from the sarcoplasmic reticulum. The Ca binds troponin, which permits the actin-myosin interaction. Vascular smooth muscles are activated by norepinephrine, which initiates a second-messenger cascade including IP_3, which triggers the release of Ca from the sarcoplasmic reticulum. The Ca binds calmodulin and the Ca-CaM complex activates myosin light-chain kinase, which phosphorylates the regulatory light chain, thus permitting the actin-myosin interaction.

Practice Examination

For each question, choose the *best* answer.

1. Hair cells are the sensory receptor cells in the cochlea. They are excited by the vibration of the hair bundle. Vibration of the hair bundle causes which one of the following events?
   a. Influx of K^+ through mechanosensitive cation channels in the tips of the cilia.
   b. Influx of Ca^{2+} through cyclic nucleotide–gated (CNG) channels in the tips of the cilia.
   c. Long-lasting hyperpolarization of the hair cell.
   d. A train of action potentials propagated from the cilia to the cell body of the hair cell.
2. Cell membranes
   a. consist almost entirely of protein molecules.
   b. are impermeable to fat-soluble substances.
   c. contain amphipathic phospholipid molecules.
   d. are freely permeable to electrolytes but not to proteins.
   e. have a stable composition throughout the life of the cell.
3. In an intestinal epithelial cell, glucose transport from the intestinal lumen to the blood involves which of the following processes?
   a. Secondary active transport
   b. Facilitated diffusion
   c. Active transport
   d. Secondary active transport and facilitated diffusion
   e. Active transport and secondary active transport

The following five choices apply to questions 4 to 7. Choices can be used more than once or not at all.

a. Acetylcholine
b. G proteins
c. GABA
d. Glutamate
e. Nitric oxide

4. This carries information from an agonist activated receptor to an indirectly activated ion channel.
5. The neurotransmitter at the neuromuscular junction.

6. The major excitatory transmitter in the CNS.
7. The major inhibitory transmitter in the CNS.
8. If all the Na-K pumps in the membrane of a muscle cell were stopped, all of the following changes would be expected for the muscle cell *except*
   a. immediate loss of the ability of the cell to carry action potentials.
   b. gradual decrease in internal K^+ concentration.
   c. gradual increase in internal Na^+ concentration.
   d. gradual decrease in resting membrane potential (the potential would become less negative).
   e. gradual increase in internal Cl^- concentration.
9. Select the one correct answer concerning ion channels:
   a. Most ion channels are open 100 percent of the time.
   b. Na^+ ions pass more readily through chloride channels than Cl^- ions do.
   c. Most ion channels are composed of subunits.
   d. A change in voltage across the cell membrane can open anion channels but never cation channels.
   e. If an ion channel carries Na^+ ion into the cell, the channel always pumps K^+ out of the cell.
10. Ca^{2+} ions are needed in the extracellular solution for synaptic transmission because:
    a. Ca^{2+} ions enter the presynaptic nerve terminal with depolarization and trigger synaptic vesicles to release their contents into the synaptic cleft.
    b. Ca^{2+} ions are required to activate glycogen metabolism in the presynaptic cell.
    c. Ca^{2+} ions must enter the postsynaptic cell to depolarize it.
    d. Ca^{2+} ions prevent Mg^{2+} ions from releasing the transmitter in the absence of nerve impulses.
    e. Ca^{2+} ions inhibit the acetylcholine esterase, enabling the released acetylcholine to reach the postsynaptic membrane.
11. Inhibitory postsynaptic potentials can arise from all of the following *except*
    a. increased permeability of the nerve membrane to Cl^- ion.
    b. direct application of GABA to neurons.
    c. increased permeability of the nerve membrane to K^+ ion.
    d. increased permeability of the cell membrane to Na^+ ion.
12. Select the correct answer. Electrical and chemical synapses differ in that
    a. Electrical synapses have a longer synaptic delay than chemical synapses.
    b. Chemical synapses can amplify a signal while electrical synapses cannot.
    c. Chemical synapses do not have a synaptic cleft while electrical synapses do have a synaptic cleft.
    d. Electrical synapses use agonist-activated channels and chemical synapses do not.
    e. Electrical synapses are found only in invertebrate animals while chemical synapses are found in all animals.

13. Which one of the following does *not* contribute to the integration of synaptic potentials by neurons?
    a. Convergence of many synaptic inputs on one neuron, allowing spatial summation.
    b. The presence of EPSPs having amplitudes that exceed the threshold for generation of an action potential in the neuron.
    c. Temporal summation of synaptic potentials in neurons due to the time constant of the neurons.
    d. The flow of currents from the distal regions of the dendrites to the soma due to the length constants of the dendrites.
    e. Inhibitory synaptic inputs.
14. Which one of the following statements is correct about the activation of different types of muscle?
    a. Mature skeletal muscle cells can be activated by one or more motor neurons.
    b. Neurons are not involved in the activation of smooth muscle cells.
    c. Cardiac muscle contraction is triggered by motor neuron activity.
    d. Postsynaptic potentials from autonomic neurons can alter skeletal muscle contraction.
    e. Autonomic neurons can alter the frequency and strength of smooth muscle contraction.
15. Energy for skeletal muscle contraction is derived from stores of which of the following?
    a. ATP, creatine phosphate, myoglobin
    b. ATP, creatine phosphate, glycogen
    c. ATP, creatine phosphate, amino acids
    d. ATP, creatine phosphate, collagen
    e. none of the above
16. Malignant hyperthermia (MH) is a disorder that afflicts a fraction of surgery patients upon exposure to certain anesthetics. The rapid high fever arises from
    a. maintained calcium release from sarcoplasmic reticulum.
    b. persistent sodium channel inactivation in the t-tubule membrane.
    c. diminished activity of the sarcolemmal Na/K-ATPase.
    d. attachment of the contractile apparatus to dystrophin.
17. Which one of the following statements concerning gap junction channels is *false*?
    a. They allow the passage of second messengers from cell to cell.
    b. They allow voltage changes in one cell to spread into other cells.
    c. They can contain one or more types of subunit.
    d. They typically are open to the extracellular space.
    e. They can be gated by voltage.
18. Which of the following ions is countertransported to energize neurotransmitter transport into presynaptic vesicles?
    a. Na^+
    b. K^+
    c. H^+

d. Cl^-
e. Ca^{2+}

19. Hyperkalemia (high extracellular potassium concentration) can stop the heart because
a. potassium ions bind to sodium channels, preventing their activity.
b. potassium ions stimulate the sodium-potassium pump and thereby prevent cardiac action potentials.
c. the membrane potential of heart cells depolarizes and its sodium channels inactivate.
d. potassium ions rush out through the inward rectifier.
e. potassium ions block the actin-myosin interaction in the heart.

20. Myelination of axons
a. reduces conduction velocity to provide more reliable transmission.
b. forces the nerve impulse to jump from node to node.
c. occurs in excess in multiple sclerosis (MS).
d. leads to a increase in effective membrane capacitance.
e. decreases the length constant for the passive spread of membrane potential.

21. Consider the following three channels in ventricular muscle cells: sodium channel (Na_V), inward rectifier potassium channel (K_{ir}), and calcium channel (Ca_V). Choose the answer that best describes which of these channels are open during the plateau phase of the ventricular action potential.
a. All three
b. Na_V and K_{ir} only
c. Ca_V and K_{ir} only
d. K_{ir} only
e. Ca_V only

22. Choose the correct statement. There is an inward current (I_f) associated with pacemaker activity in cells of the sinoatrial node. Stimulation of sympathetic nerves leading to the heart or application of norepinephrine produces
a. a decrease of I_f, a decrease in heart rate, and an increase in strength of contraction.
b. a decrease of I_f, an increase in heart rate, and an increase in strength of contraction.
c. an increase of I_f, an increase in heart rate, and an increase in strength of contraction.
d. an increase of I_f, a decrease in heart rate, and a decrease in strength of contraction.
e. an increase of I_f, an increase in heart rate. and a decrease in strength of contraction.

23. A solution is prepared by adding 10 g of NaCl (formula weight = 58.5) to 1 L of distilled water. An isotonic solution is 300 mosm. The prepared solution is
a. very hypotonic (with less than 50% normal tonicity).
b. slightly hypotonic (about 10% low).

c. isotonic (within 1%).
d. slightly hypertonic (about 10% high)
e. very hypertonic (more than twice normal tonicity).

24. Drinking isotonic saline solution will decrease
a. extracellular volume.
b. extracellular osmolarity.
c. intracellular volume.
d. intracellular osmolarity.
e. none of the above.

25. A 34-year-old man develops a herpes infection of the cornea, a major cause of infectious corneal blindness. The virus itself replicates in the trigeminal ganglion in sensory neurons innervating the cornea. Which one of the following was the most likely route by which those neurons became infected in the first place?
a. Virus on the cornea was taken up by nerve terminals and orthogradely transported to the cell body.
b. Virus on the cornea was taken up by nerve terminals and retrogradely transported to the cell body.
c. Virus on the lips was taken up by nerve terminals and retrogradely transported to the cell body.
d. Virus on the lips was taken up by nerve terminals and orthogradely transported to the cell body.
e. Virus inhaled in droplets entered the bloodstream and traveled to the neurons in the trigeminal ganglion.

26. A branch of a 26-year-old man's ulnar nerve was crushed in his left forearm, severing axons at a point about 200 mm from the skin on the medial part of the palm, where cutaneous sensation was lost. About how long will it likely take before the patient begins to feel stimuli in that part of the palm?
a. 1 day
b. 10 days
c. 100 days
d. 1000 days
e. never, since peripheral axons do not regenerate

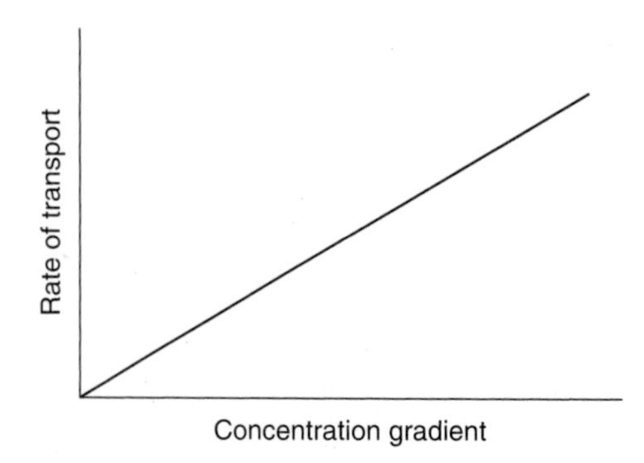

27. The diagram above is typical for the concentration gradient dependence of
 a. the rate of secondary active transport.
 b. the rate of primary active transport.
 c. the rate of transport by passive diffusion.
 d. the rate of transport by facilitated diffusion.

28. Which one of the following is *least likely* to regulate Na^+/K^+ pump activity?
 a. cardiac glycosides
 b. second messengers (e.g., cAMP, diacylglycerol)
 c. intracellular Na^+ concentration
 d. extracellular Mg^{2+} concentration
 e. extracellular K^+ concentration

29. Mark the *false* statement
 Electrical synapses
 a. can rectify.
 b. are gap junctions in the nervous system.
 c. have a longer synaptic delay than chemical synapses.
 d. do not require transmitters.
 e. provide direct electrical continuity between neurons.

30. Treatments for nerve gas poisoning target which of the following proteins?
 a. Acetylcholinesterase (AChE) and choline acetyltransferase (CAT).
 b. AChE and nicotinic acetylcholine receptors.
 c. Muscarinic and nicotinic acetylcholine receptors.
 d. Muscarinic acetylcholine receptors and AChE.
 e. CAT and synaptic choline transporters.

31. The major pathway of degradation of glutamate in glial cells is catalyzed by
 a. glutamine synthetase.
 b. glutaminase.
 c. tyrosine hydroxylase.
 d. GABA transaminase.
 e. glutamate decarboxylase.

32. Which of the following values is the closest representation of the concentration of serotonin in presynaptic vesicles?
 a. 50 pM
 b. 50 nM
 c. 50 μM
 d. 50 mM
 e. 50 M

33. Negative feedback control systems do *not*
 a. improve the reliability of control.
 b. require the sensing or measurement of the controlled process.
 c. require communication between separate parts of the system.
 d. regulate blood pressure and body temperature.
 e. cause the all-or-none property of the action potential.

34. Propagation of a nerve impulse does *not* require
   a. closure of potassium channels that maintain the resting potential.
   b. a conformational change in membrane proteins.
   c. a membrane depolarization that opens Na channels.
   d. current to enter the axon and flow within the axon.
   e. entry of sodium ions into the axon.

35. Voltage-clamp experiments on nerve membranes
   a. mechanically squeeze the nerve and measure the voltage.
   b. electronically control the membrane potential and measure the current through the membrane.
   c. hold the current through the membrane constant and measure changes of the membrane potential.
   d. electronically control both the membrane potential and the membrane current and measure the mechanical squeezing produced by the cell.
   e. are a theoretical concept and have never been physically performed.

36. The compound action potential recorded with a pair extracellular electrodes from an intact bundle of nerve fibers
   a. propagates without change in size or shape.
   b. is all-or-none. If a threshold is exceeded, further increase in stimulus does not increase the response.
   c. has an amplitude of about 100 mV.
   d. is biphasic, showing both upward and downward deflections from the baseline.
   e. is *not* blocked by tetrodotoxin (TTX).

37. A scientist is recording from the soma of a neuron with an intracellular microelectrode to study synaptic inputs on the dendrites. The letters a, b, and c below indicate the synaptic potentials recorded from three different synaptic inputs. For identical synaptic inputs to the dendrites, which synaptic potential was generated by the synapse at a location on the dendrites closest to the soma?

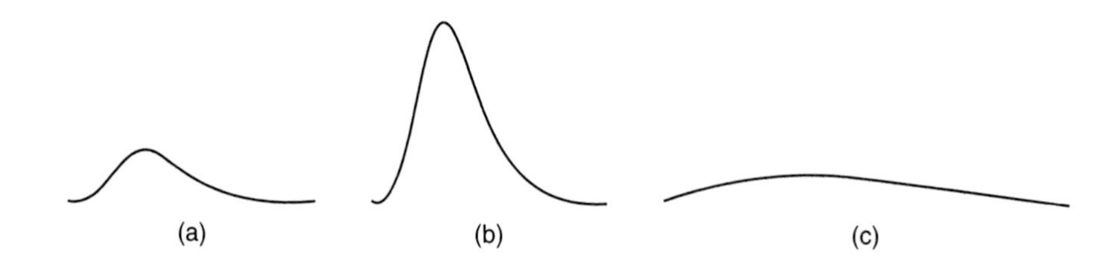

38. If the potassium ion concentration on the outside of a resting skeletal muscle cell is reduced to half of the normal value by removing K^+ and Cl^-

in equal amounts, what would be the best estimate of the effect on the resting membrane potential?

a. Hyperpolarize about 100 mV.
b. Depolarize about 5 mV.
c. Hyperpolarize about 15 mV.
d. Depolarize about 20 mV.
e. No measurable effect.

39. The following cell in an organism called the *Europa* louse was recovered from a moon of Jupiter with a space probe. The intracellular and extracellular concentrations of all the ions are given below:

Extracellular	Intracellular
Rb^+ = 100 mM	Rb^+ = 1 mM
SO_4^{2-} = 50 mM	SO_4^{2-} = 0.5 mM

The cell membrane is permeable to Rb^+ and not to SO_4^{2-} or water.

What is the resting membrane potential? (The sign refers to the potential inside of the cell.)

a. +30 mV
b. +60 mV
c. +120 mV
d. −30 mV
e. −60 mV

40. For neuromuscular transmission at skeletal muscles, which one of the following is true?

a. The transmitter that is released is not stored in synaptic vesicles before release.
b. There is a time delay between the depolarization of the presynaptic nerve terminal and the generation of the postsynaptic endplate potential.
c. The postsynaptic response is inhibitory only.
d. The neuromuscular junction transmits electrical potentials in two directions: nerve to muscle and muscle to nerve.
e. A chemical transmitter substance is not released.

41. A patient's arm muscles become progressively weaker during repeated lifting of a weight. Clinical nerve conduction tests indicate that repetitive stimulation of the axons innervating the involved muscles produces compound action potentials that do not change in amplitude during the repetitive stimulation. Direct stimulation of the muscle with needle electrodes produces muscle action potentials and muscle contractions that do not decrease in strength during repetitive stimulation.

Which one of the following is the most likely cause of the muscle weakness?

a. A depletion of ATP in the muscle cells.
b. A defect in the propagation of action potentials in axons.
c. A defect in the propagation of action potentials in muscle.
d. A defect in neuromuscular transmission.
e. A defect in the contractile mechanism of muscle.

42. A drug competes with acetylcholine for the acetylcholine binding site on the enzyme acetylcholinesterase. It is expected that moderate doses of this drug would

a. decrease the amplitude of the endplate potential.
b. increase the amplitude of the endplate potential.
c. have no effect on the amplitude of the endplate potential.
d. increase the rate at which acetylcholine is hydrolyzed in the synaptic cleft.

43. The cellular mechanism for terminating muscle contraction is which one of the following?

a. Dephosphorylation of troponin
b. Depletion of ATP
c. Sequestration of calcium
d. Activation of voltage-gated calcium channels
e. Activation of sodium-potassium ATPase

44. Which one of the following statements is correct about contractile filaments in muscle?

a. Actin and myosin filaments are found in skeletal and cardiac but not smooth muscle.
b. Tropomyosin molecules are part of the thick filament in smooth muscle.
c. Sarcomeres in smooth muscle are smaller than those in striated muscle.
d. Calcium regulates striated muscle by binding to proteins associated with thin filaments.
e. Calcium regulates smooth muscle by binding to proteins associated with thin filaments.

45. In the length-tension characteristics of skeletal muscle, which one of the following is true?

a. The active tension curve is the same as the passive tension curve.
b. Passive tension is attributed to the ATP content of muscle.
c. Passive tension curves are identical when different muscles are tested.
d. Maximal active tension is generated when thin and thick filaments have minimum overlap.
e. None of the above is true.

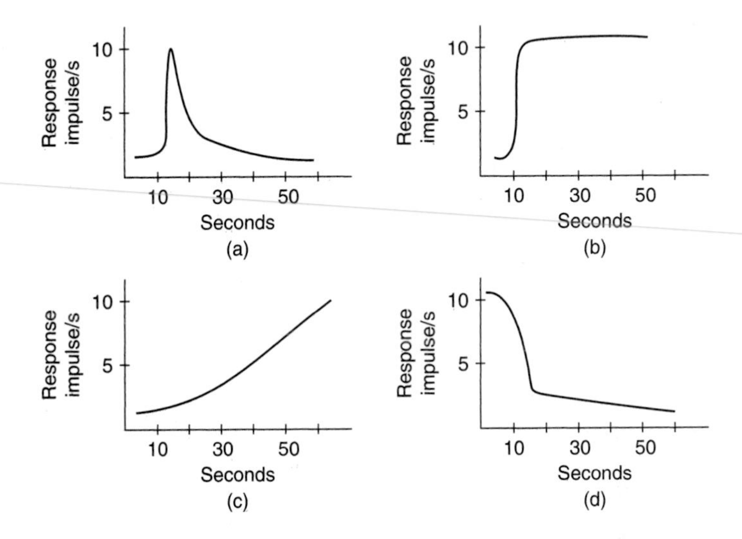

46. The graphs above show the frequency of action potentials (y axis) recorded from a primary sensory afferent fiber during sensory stimulation. Which one of these graphs shows the response from a typical sensory fiber (excluding pain fibers) to a *constant* maintained stimulus applied beginning at 10 s and lasting throughout the recording (i.e., until 50 s)?

Answers to Practice Examination

1.	a	17.	d	33.	e
2.	c	18.	c	34.	a
3.	d	19.	c	35.	b
4.	b	20.	b	36.	d
5.	a	21.	e	37.	b
6.	d	22.	c	38.	c
7.	c	23.	d	39.	c
8.	a	24.	e	40.	b
9.	c	25.	b	41.	d
10.	a	26.	c	42.	b
11.	d	27.	c	43.	c
12.	b	28.	d	44.	d
13.	b	29.	c	45.	e
14.	e	30.	d	46.	a
15.	b	31.	a		
16.	a	32.	d		

Index

NOTE: Page numbers followed by *f* indicate figures; those followed by *t* indicate tables.